DATE DUE

NOV 0 6 1998			
APR 1 6 2000			
MAY 1 5 2001			
DEC 1 2 2001			
DEC 0 4 2009			
MAR 0 5 2012			

HIGHSMITH 45-220

Wastewater Disinfection

Manual of Practice FD-10

Prepared by **Task Force on Wastewater Disinfection**
Domenic Grasso, *Chair*

Barnes R. Bierck
Ernest R. Blatchley III
Gerald F. Connell
Laxman Mani Devkota
Gordon R. Finch
Michael Fleury
Joseph A. Guagno, Sr.
Frederick L. Hart
Victor Hom
Samuel S. Jeyanayagam

Nancy E. Kinner
Shundar Lin
Manu Patel
Anne M. Penkal
Otto K. Scheible
Lynn H. Teuscher
John Throne
Rufino A. Vigilia, Jr.
Robert J. Walker
Larry W. Zimmerman

Under the Direction of the
**Municipal Subcommittee of the
Technical Practice Committee**

1996

**Water Environment Federation
601 Wythe Street
Alexandria, VA 22314-1994 USA**

Library of Congress Cataloging-in-Publication Data

Wastewater disinfection/prepared by Task Force on Wastewater Disinfection
under the direction of the Municipal Subcommittee of the Technical Practice
Committee.

    p. cm. — (Manual of practice : FD -10)

    Includes bibliographical references and index.

    ISBN 1-57278-036-3 (hardcover)

    1. Sewage — Purification. I. Water Pollution Control Federation.
Task Force on Wastewater Disinfection. II. Water Environment Federation.
Municipal Subcommittee. III. Series: Manual of Practice. FD; no 10.

TD747.W37 1996

628.3'2—dc20

Copyright ©1996 by the Water Environment Federation
Alexandria, VA 22314-1994 USA

Library of Congress Catalog Card No. (96-24225)
ISBN 1-57278-036-3
Printed in the USA **1996**

Water Environment Federation

The Water Environment Federation is a not-for-profit technical and educational organization that was founded in 1928. Its mission is to preserve and enhance the global water environment. Federation members are more than 42,000 water quality specialists from around the world, including environmental, civil, and chemical engineers; biologists; chemists; government officials; treatment plant managers and operators; laboratory technicians; college professors, researchers, and students; and equipment manufacturers and distributors.

For information on membership, publications, and conferences, contact

Water Environment Federation
601 Wythe Street
Alexandria, VA 22314-1994 USA
(703) 684-2400

Manuals of Practice for Water Pollution Control

The Water Environment Federation (WEF) Technical Practice Committee (formerly the Committee on Sewage and Industrial Wastes Practice of the Federation of Sewage and Industrial Wastes Associations) was created by the Federation Board of Control on October 11, 1941. The primary function of the committee is to originate and produce, through appropriate subcommittees, special publications dealing with technical aspects of the broad interests of the Federation. These manuals are intended to provide background information through a review of technical practices and detailed procedures that research and experience have shown to be functional and practical.

IMPORTANT NOTICE

The contents of this publication are for general information only and are not intended to be a standard of the Water Environment Federation (WEF).

No reference made in this publication to any specific method, product, process, or service constitutes or implies an endorsement, recommendation, or warranty thereof by the Federation.

The Federation makes no representation or warranty of any kind, whether expressed or implied, concerning the accuracy, product, or process discussed in this publication and assumes no liability.

Anyone using this information assumes all liability arising from such use, including but not limited to infringement of any patent or patents.

Acknowledgments

This manual was produced under the direction of Domenic Grasso, *Task Force Chair*. The principal authors are

Barnes R. Bierck
Ernest R. Blatchley III
Gerald F. Connell
Audrey A. Cudrak
Laxman Mani Devkota
Gordon R. Finch

Michael Fleury
Frederick L. Hart
Victor Hom
Samuel S. Jeyanayagam
Nancy E. Kinner
Otto K. Scheible

Additional information and review were provided by John S. Maynes.

Authors' and reviewers' efforts were supported by the following organizations:

Bailey, Fischer & Porter Co., Warminster, Pennsylvania
Camp Dresser & McKee, Inc., Tucson, Arizona
Capital Controls Company, Inc., Colmar, Pennsylvania
Damon S. Williams Associates, L.L.C., Phoenix, Arizona
EMA, Inc., Philadelphia, Pennsylvania
George Washington University, Washington, D.C.
HDR Engineering Inc., Bellevue, Washington
HydroQual, Inc., Mahwah, New Jersey
IIASA, Germany
Illinois State Water Survey, Peoria
Lewis & Zimmerman Associates, Rockville, Maryland
Purdue University, West Lafayette, Indiana
Stanley Associates, Surrey, British Columbia, Canada
Team Environmental Services, San Marcos, California
University of Alberta, Edmonton, Canada
University of Connecticut, Storrs
University of New Hampshire, Durham
West Boise Wastewater Plant, Boise, Idaho
Worcester Polytechnic Institute, Worcester, Massachusetts

Federation technical staff project management was provided by Christine Koppel; technical editorial assistance was provided by Matthew Hauber.

Figure Contributors

Bailey, Fischer & Porter Ltd.   7.17, 7.18
   Warminster, Pennsylvania

Capital Controls Company, Inc.   5.10, 5.11, 5.12, 5.16, 5.17
   Colmar, Pennsylvania

Infilco Degremont, Inc.   7.1(b), 7.16, 7.20
   Richmond, Virginia

Light Sources, Inc.   7.21

Trojan Technologies, Inc.   7.1(a)
   London, Ontario, Canada

Contents

Chapter		Page
1	Introduction	1
2	Regulatory and Public Health Concerns	5
	Introduction	5
	Public Health Concerns	7
	Regulatory Concerns	9
	National	9
	State	10
	Canadian Regulatory Concerns	12
	Introduction	12
	Regulatory Framework	14
	Regulations and Guidelines	14
	Disinfection Requirements	15
	References	17
3	Microbial Fundamentals	19
	The Need for Wastewater Disinfection	20
	Introduction	20
	Water Quality Standards and Disinfection Criteria	20
	History of Waterborne Disease Outbreaks	21
	Drinking Water	21
	Recreational Water	26
	Shellfish	27
	The Mills–Reincke Phenomenon and the Hazen Theorem	27
	Need for Disinfection—Summary	28
	Pathogens, Diseases, and Microbial Fundamentals	28
	Common Epidemiological Terms	28
	Spread of Infections and Preventive Measures	29
	Spread of Infections	29
	Preventive Measures	30
	Immunization	30
	Principal Pathogens of Concern	30
	Basic Types of Pathogenic Organisms	30
	Bacteria	31
	Fungi	31
	Protozoa	31

Algae		31
Viruses		32
Pathogens and Disease		32
Pathogens and Indicator Organisms		37
Survival of Pathogens in the Environment		39
Wastewater Disinfection Technologies		40
Types of Disinfection Technologies		40
Disinfection Technologies and Mechanisms of Microbial Inactivation		45
Chlorine		45
Ozone		45
Ultraviolet Radiation		46
Microbial Inactivation Unrelated to Disinfection		47
Disinfection Kinetics		48
Dose, Residual, Demand, and Contact Time		48
Kinetic Models of Disinfection		48
Regrowth Phenomenon		51
Plant Control of Disinfection		51
Residual Toxicity of Disinfectants		52
References		54
4	Reactor Dynamics	61
Introduction		62
Ideal Reactors		62
Tracer Test Methods		65
Wastewater Flow Conditions		65
Tracer Selection		66
Tracer Injection and Detection		67
Tracer Curves		69
E Curve		69
C Curve		71
F Curve		71
Mean Residence Time		73
Variance		74
Nonideal Reactors		75
Typical Wastewater Disinfection Reactors		75
Conceptual Models of Nonideal Reactors		76
The Segregation Model		77
The Dispersion Model		77
The Tanks-in-Series Model		79

Tracer Data Interpretation   80

    The d and N Indices   80
    Other Index Values   80

Tracer Analysis Examples   82

    Example 1: Step-Feed Test   82
        Data Collection   82
        Tracer Results   82
        Data Analysis   83
        Data Interpretation   84
    Example 2: Pulse-Feed Test   84
        Data Collection   84
        Tracer Results   87
        Data Analysis   87
        Data Interpretation   87

Combining Tracer Analysis and Disinfection Kinetics   88

    Disinfection Kinetics   90
    Ideal Models   91
    Using the Segregated Model   91
    Using the Tanks-in-Series Model   92
    Using the Dispersion Model   92

Calculating a Reactor's Potential Efficiency   92

    Example 3: Disinfection Efficiency Estimation   93
        Data Collection   93
        Tracer Results   93
        Data Analysis   93
        Data Interpretation   93
        Other Criteria That Influence Disinfection Efficiencies 93

Reactor Design Considerations   97

    Baffles   97
    Configuration   97
    Surrounding Conditions   97

References   98

Suggested Readings   99

5   Chlorination/Dechlorination   101

Chemistry of Chlorine and Sulfur   102

    Elemental Chlorine   102
        Physical Properties   102
        Chemical Properties   104
        Toxicity   104
    Hypochlorites   104
        Physical Properties   105

Chemical Properties	105
Sulfur Dioxide	105
Physical Properties	105
Chemical Properties	106
Toxicity	106
Sulfite Salts	106
Chemistry and Reactions	106
Chlorine	106
Chlorine Dioxide	107
Inorganic Reactions	109
Chloramines	109
Breakpoint	110
Organic Reactions	111
Disinfection	113
Chlorine Toxicity and Effects on Higher Organisms	113
Aftergrowths	113
Free Versus Combined Chlorine Residual	114
Reduction of Chlorine Residuals	116
Dechlorination Reactions	116
Safety and Health	117
Chlorine Gas	117
Hypochlorites	118
Sulfur Dioxide	118
Shipment and Handling Safety	119
Cylinders	119
Containers	121
Facility Design	121
Ton Containers	124
Vaporizer Facilities	124
Container Hookup	125
Analytical Determination of Chlorine Residuals	128
Methods	128
Iodometric	128
Starch-Iodide	128
Amperometric	129
N,N-Diethyl-*p*-Phenylene-Diamine	129
Titration	129
Colorimetry	129
Leuco Crystal Violet	129
Free Available Chlorine Test Syringaldazin	130
Note on Orthotolidine	130
Selection of Method	130

Process Design Requirements 131

    Disinfection Factors 131
    Mixing 131
    Closed Conduits 132
    Hydraulic Devices 133
    Contacting 133

Design and Selection of Equipment 135

    Chlorinators/Sulfonators 135
    Chemical-Feed Pumps 136
    Sulfur Dioxide Feeders 138
    Manifolds and Vacuum Regulator Location 139
    Vaporizers 140
    Residual Analyzers 142
    Maintenance 143

Feed Control Strategies 144

    Manual Control 145
    Semiautomatic Control 145
    Flow-Proportional Control 146
    Residual Control 147
    Compound-Loop Control 149
    Cascade Control 150
    Manual Injection Points 150
    Dechlorination Control 153

References 155

6   Ozone Disinfection 157

History 158

Chemistry 159

    Physical and Chemical Properties 159
    Ozone Chemistry in Aqueous Solution 161
      Ozone Decomposition 161
      Reaction Pathways and the Fate of Ozone 161
    Measuring Ozone Concentration 165
      Gas-Phase Concentration Measurement 165
      Aqueous-Phase Concentration Measurement 167
    Toxicological Properties 168
      Occupational Exposure 168
      Residual Toxicity in the Water Environment 170

Disinfection 172

    Mechanisms 172
    Factors Affecting Disinfection Efficacy 173
      Effect on Bacteria 175
      Effect on Viruses 175

Effect on Protozoa   177

Design   179

  Feasibility   179
    Bench-Scale Tests   180
    Pilot-Scale Tests   180
  Process Components   181
  Feed Gas Selection and Preparation   181
    Source Gas   181
    Gas Treatment Systems   182
  Ozone Generation   184
    Theory   184
    Types of Generators   185
    Cooling   186
  Contacting   188
    Objectives   188
      Gas Mass Transfer   189
      Disinfection Efficiency   190
    Types of Contactors   191
    Contactor Design Considerations   195
      Fine-Bubble Diffusers   195
      Injector and Static Mixer   197
      Turbines   197
    Process Control   197
  Offgas Destruction   198
  Materials   198
  Process Design Example   198
    Step 1: Determine the Initial Ozone Demand and Slope
      of the Dose-Response Curve   202
    Step 2: Choose the Ozone-Transfer Efficiency and Size
      the Contactor   203
    Step 3: Determine the Design Ozone Production Rate   203
    Step 4: Select the Number of Ozone Generators   203
    Step 5: Determine the Size and Number of Air
      Compressors   205
    Step 6: Determine the Size and Number of Desiccant
      Dryers   206

Operation, Maintenance, and Safety   208

  Operation   208
    General   208
    Gas Preparation   209
    Electrical Power Supply   210
    Ozone Generator   210
    Ozone Contactor   210
    Offgas-Destruction System   210
    Additional Considerations   210

Maintenance 211
General 211
Gas Preparation 211
Electrical Power Supply 212
Ozone Generator 212
Ozone Contactor 212
Associated Equipment 213
Safety 213

References 215

7   Ultraviolet Disinfection 227

Introduction 227

General Description of Ultraviolet Disinfection 230

Photoreactivation and Dark Repair 233

Ultraviolet Inactivation Kinetics 236

Other Kinetic Models 244

Effect of Intensity on Inactivation Behavior 248

Intensity 249

Ultraviolet Dose 257

Hydraulics 259
Longitudinal Dispersion 259
Head Loss 264

Factors Affecting Lamp Output 267

Mathematical Models 269

Fouling 273

General Considerations in Ultraviolet System Design 275
Design Wastewater Characteristics 276
Pilot Testing 279
System Sizing and Configuration Considerations 280
Retrofit Considerations 282

Current Ultraviolet Equipment 282
Low-Pressure Mercury Lamp Systems 283
Horizontal Ultraviolet Systems 284
Vertical Ultraviolet Systems 285
Medium-Pressure Mercury Lamp Systems 286
Low-Pressure, High-Intensity Systems 288

References 289

Index 293

List of Tables

Table		Page
2.1	Typical wastewater influent concentration ranges for pathogenic and indicator organisms	7
2.2	Waterborne diseases and transmission routes	8
2.3	Secondary effluent ranges for pathogenic and indicator organisms before disinfection	9
2.4	Canadian legislative and regulatory documents and agencies	13
2.5	Maximum coliform limits in Canada	15
3.1	Waterborne disease outbreaks, 1971 to 1980, by type of illness and water system	22
3.2	Waterborne disease outbreaks related to drinking water during 1980 and 1981	23
3.3	Waterborne disease outbreaks related to drinking water during 1991 and 1992	24
3.4	Disease outbreaks related to recreational water during 1980 and 1981	26
3.5	Routes of infection	29
3.6	Summary of taxonomic, clinical, and epidemiological features of potential drinking water disease agents	33
3.7	Bacterial densities in domestic wastewater	38
3.8	Typical levels of coliform bacteria in domestic wastewater after various wastewater treatment steps	39
3.9	Microbial reductions by conventional treatment processes	39
3.10	Survival times of pathogens in soil and on plant surfaces	40
3.11	Major factors in evaluating disinfectant alternatives	42

3.12	Applicability of alternative disinfection techniques	43
4.1	Typically used tracers	67
5.1	Expected virus concentration in effluent	114
5.2	Chlorine design requirements to disinfect normal domestic wastewater as listed in various references	132
6.1	Selected properties of pure ozone	160
6.2	Ozone concentration relationships at standard temperature and pressure (standard temperature and pressure are 273.15°K and 101.33 kPa)	160
6.3	Health effects of ozone	168
6.4	Effects of residual ozone on aquatic organisms	171
6.5	Summary of reported ozonation requirements for inactivation of *Cryptosporidium* sp. oocysts compared with published requirements for *Giardia lamblia*	178
6.6	Characteristics of tube-type generators	187
6.7	Summary of gas–liquid contacting systems	192
6.8	Materials resistant to corrosion	199
6.9	Ozone disinfection system criteria for example design problem	201
6.10	Summary of dose-response data collected for ozone disinfection of *E. coli* in a nitrified secondary effluent	201
6.11	Design example calculations	204
6.12	Moisture content of air for air temperatures from −80 to 40°C	207
7.1	Bond energies of importance in microbiological systems	232
7.2	Reported microbial dose-response behavior resulting from ultraviolet irradiation	240
7.3	Multihit and series-event model parameters for inactivation kinetics	246

List of Figures

Figures		Page
3.1	Incidence of waterborne disease from 1971 to 1992 as a function of water system	25
4.1	Plug-flow reactor	63
4.2	Continuously stirred tank reactor	63
4.3	Plug-flow reactor showing step-feed input and F curve output, and slug-injection input and E curve output	64
4.4	Continuously stirred tank reactor showing step-feed input and F curve output, and slug-injection input and E curve output	65
4.5	Calculating c_i using the midpoint method	70
4.6	The E curve	71
4.7	Measured concentrations to generate F curve	72
4.8	The F curve	72
4.9	Effects of variance on the C,E curve	75
4.10	Serpentine flow reactor	76
4.11	Cross section of a typical ultraviolet reactor	77
4.12	E curve response for variations of the dispersion model	78
4.13	E and F curves for tanks-in-series model	80
4.14	Typical trace curve parameters	81
4.15	Example 1 (step feed)	83
4.16	Example 1 data output	84
4.17	Example 1 (F curve)	85

4.18	Example 1 (*E* curve)	86
4.19	Example 2 (pulse feed)	87
4.20	Example 2 data output	88
4.21	Example 2 (*E* curve)	89
4.22	Example 3 data output	94
4.23	Example 3 (*E* curve)	95
4.24	Example 3 disinfection analysis	96
5.1	Solubility of chlorine in water	103
5.2	Chlorine, hypochlorous acid: hypochlorite distribution versus pH	108
5.3	Chlorine residual with ammonia nitrogen	111
5.4	Chlorine residual with ammonia nitrogen and organic nitrogen	112
5.5	*Escherichia coli* kill times versus residual concentration	115
5.6	Types of 70-kg (150-lb) cylinders	122
5.7	Ton container orientation for proper use	123
5.8	Ton container manifold for liquid chlorine withdrawal with optional gas bypass	126
5.9	Automatic pressure switchover system	127
5.10	Schematic gas induction/mixing device	134
5.11	Schematic of cylinder-mounted, vacuum-operated gas chlorinator	136
5.12	Schematic of direct-pressure-operated gas feeder	137
5.13	Chemical-feed system schematic	138
5.14	Typical gas manifold with heaters	139

5.15	Automatic vacuum switchover system	140
5.16	Vaporizer schematic	141
5.17	Amperometric analyzer schematic	143
5.18	Flow-proportional control schematic	146
5.19	Residual control schematic	147
5.20	Compound-loop schematic	149
5.21	Cascade control schematic	151
5.22	Multiple manual injection point schematic	152
5.23	Dechlorination feed-back schematic	153
5.24	Dechlorination feed-forward schematic	154
6.1	Two mechanisms for ozone decomposition in pure water	162
6.2	Typical ozone reactions in aqueous solution	163
6.3	Human tolerance for gaseous ozone	169
6.4	Ozone disinfection of total coliforms in various process effluents	174
6.5	*E. coli* survival as a function of ozone residual in a nitrified secondary effluent	176
6.6	Relative sensitivity of MS2 coliphage and poliovirus type 3	177
6.7	Relative susceptibility of MS2, *E. coli, Giardia, Cryptosporidium,* and heterotrophic plate count bacteria to ozone	179
6.8	Flow diagram for ozone wastewater disinfection	181
6.9	Components of low-pressure air feed gas treatment schematic	183
6.10	Ozone production versus air dew point	184

6.11   Cross-sectional view of principal elements of a corona discharge
       ozone generator   185

6.12   Decomposition of ozone in air   188

6.13   Schematic of a four-stage bubble diffuser ozone contact basin
       (for low gas-phase ozone concentrations)   191

6.14   Schematic of a gas induction system with in-line static mixer
       (for high gas-phase ozone concentrations)   195

6.15   Schematic of a submerged turbine mixer ozone contactor
       (for low through high gas-phase ozone concentrations)   196

6.16   Sample dose-response curve for ozone disinfection of *E. coli* in
       secondary effluent   202

7.1   Schematic illustration of open-channel ultraviolet disinfection
       system with horizontal lamp configuration, and schematic
       illustration of open-channel ultraviolet disinfection system with
       vertical lamp configuration   229

7.2   Radiant power output spectra from low-pressure and
       medium-pressure mercury arc lamps   231

7.3   Ultraviolet absorption spectra for purine and pyrimidine bases
       at pH = 7   232

7.4   Hypothesized photoreactivation reaction mechanism   234

7.5   Schematic illustration of collimation apparatus for ultraviolet
       exposure experiments   237

7.6   Observed deviations from ideal first-order dose-response
       behavior resulting from ultraviolet irradiation   242

7.7   Effect of filtration on coliform survival in wastewater effluent   242

7.8   Effect of filtration on particle size distribution of secondary
       effluent from a wastewater treatment facility   243

7.9   Application of multitarget kinetic model to inactivation response
       of *E. coli*, *C. parapsilosis*, and coliphage f2 in batch reactors 246

7.10   Application of series-event kinetic model to inactivation response
        of *E. coli*, *C. parapsilosis*, and coliphage f2 in batch reactors   247

7.11   Schematic illustration of lamp geometry as used in point-source
        summation technique   251

7.12   Intensity distributions calculated by point-source summation for
        147-cm (1 470-mm) arc length low-pressure mercury arc lamp   253

7.13   Intensity distributions at a radial distance of 1.0 cm (10 mm) from
        the outside of quartz lamp jackets for low-pressure mercury arc
        lamps used in open-channel ultraviolet disinfection systems   254

7.14   Intensity fields calculated by point-source summation within an
        array of four lamps with axes at the corners of a 7.6 cm × 7.6 cm
        square   255

7.15   Schematic representation of bioassay procedure   257

7.16   Profile schematic of lamp modules relative to inlet and outlet
        structures   261

7.17   Schematic illustration of ultraviolet disinfection system with
        stilling plate for flow conditioning and elongated weir for level
        control   262

7.18   Schematic illustration of flap gate and submerged dam used as
        outlet structure at Bonnybrook wastewater treatment plant,
        Calgary, Alberta, Canada   263

7.19   Measured values of head loss (per bank) as a function of
        approach velocity in open-channel ultraviolet disinfection
        systems   265

7.20   Use of a stepped channel to minimize the effects of head loss
        in an open-channel ultraviolet disinfection system   266

7.21   Typical ultraviolet lamp output as a function of time   268

7.22   Determination of the water quality factor (f) by graphical means   272

7.23   Estimation of effluent coliform viability (*N*) by graphical means   272

7.24   Model predictions of effluent viability for *E. coli*, *C. parapsilosis*, and coliphage f2 from a complete-mix ultraviolet reactor using the multitarget and series-event inactivation models   274

7.25   Regression of observed effluent fecal coliform concentration versus suspended solids concentration compiled from several studies of ultraviolet disinfection systems   278

Chapter 1
Introduction

Among the many provisions of the Federal Water Pollution Control Act Amendments of 1972 (PL 92-500) is a unified basis for determining secondary treatment disinfection standards. As part of this comprehensive regulatory framework, disinfection standards were mandated through the establishment of fecal coliform criteria.

Fecal coliform bacteria are typically used as indicator species to ascertain the existence of other microorganisms (enteric bacteria and viruses) that may pose public health risks. Other human pathogens such as *Giardia* and *Cryptosporidium* are present in the absence of coliform indicators and may not be inactivated using standard disinfection processes. Therefore, regulations that mandate fecal coliform limits may not entirely protect public health from challenges posed by parasites that survive conventional treatment. Wastewater effluent disinfection can be accomplished through a variety of techniques, such as chlorination, ozonation, or ultraviolet irradiation.

To determine appropriate and efficacious disinfection techniques for a particular scenario, field studies to gather stream data, wastewater characteristics, and treatment train performance criteria must be conducted. Bench- and pilot-scale studies can then be conducted to acquire appropriate design parameters and predict reliability of system performance to meet regulatory requirements.

Risks associated with the decision to disinfect effluents may extend to public health and natural resource domains. Under certain circumstances, discontinuation of the disinfection process may be warranted. For example, chlorine may react with aquatic dissolved organic matter to form potentially carcinogenic halogenated byproducts. Should the receiving stream have limited uses (for example, recreation), only seasonal disinfection may be required. During the winter, when the potential for human contact is minimal, it may be prudent and economical to discontinue disinfection, reducing anthropogenic loads (for example, chlorine) on receiving bodies.

This document is a revision of the 1986 edition of Manual of Practice FD-10, *Wastewater Disinfection,* and is intended to serve as guidance for engineers, scientists, and wastewater treatment plant operators in the comparison, selection, design, and operation of various commonly employed wastewater disinfection processes. In addition to major sections on chlorination, ozonation, and ultraviolet radiation, the revised manual includes new sections on regulations, microbial fundamentals, and reactor dynamics. Significant omissions from the earlier manual include sections on chlorine dioxide and bromine chloride. Both bromine chloride and chlorine dioxide have found limited use in wastewater disinfection, as there remain large gaps in aquatic toxicity of byproducts and comparative biocidal efficacies.

The overall structure of the manual is based primarily on addressing fundamental issues that are generically extensible to a variety of applications. An understanding of comparative regulatory requirements, basic microbial principles, and reactor performance is essential to all disinfection processes. These principles are fundamental in nature and, consequently, can be drawn on continuously, even as applied technology changes and progresses. Subsequent to an overview of fundamentals, common disinfection technologies are discussed from both a fundamental scientific perspective and an engineering design and operations point of view.

Chapter 2 on regulations provides the regulatory basis for disinfection and an overview of U.S. federal and state regulations, as well as summarizing Canadian requirements and guidelines.

Chapter 3 on microbial fundamentals complements the regulations chapter by providing a scientific basis for disinfection. The microbial basis for disease is discussed, addressing epidemiological principles and various pathogens of concern. Wastewater disinfection technologies and their various mechanisms and kinetics of inactivation are also discussed. Finally, residual disinfectant toxicity is reviewed.

Disinfection contact basin performance is generally dictated by system-specific transport and reaction and inactivation processes. The impact and sensitivity of these processes and associated parameters can be determined from mass-balance formulations incorporating mathematical representations of transport phenomena and the kinetics of the operative processes. Performance of full-scale systems can generally be predicted from these formulations. However, as part of the system hydrodynamic representation, an accurate assessment of flow patterns and detention times must be developed. Chapter 4 on reactor dynamics provides a basis for evaluating contact basin performance efficiencies. The theoretical and practical basis for tracer studies is discussed.

Chapter 5 on chlorination and dechlorination is, in large measure, similar to the previous edition's section covering this material. Chlorine chemistry and practical applications are discussed. Process and facilities design considerations, including instrumentation and control, are detailed.

Chapter 6 on ozone disinfection provides a condensed and updated version of material from the preceding manual. Again, ozonation history, chemistry, performance, and public health aspects are addressed. Engineering aspects of system design and operation and maintenance (O & M) considerations for generation, dissolution, offgas treatment, and control systems are presented.

Chapter 7 discusses the use of ultraviolet radiation for disinfection. Again, this chapter has been significantly modified to address mechanisms and kinetics of disinfection, hydraulics of contactors, determination of dose or intensity, development of predictive models, system performance, photoreactivation, and design and O & M considerations.

Although extensive reference lists are provided at the end of each chapter, three excellent general sources are the following:

- Bryant, E.A. (1992) *Disinfection Alternatives for Safe Drinking Water.* Van Nostrand Reinhold, New York, N.Y.
- U.S. Environmental Protection Agency (1986) *Design Manual: Municipal Wastewater Disinfection.* EPA-625/l-86-021, Cincinnati, Ohio.
- White, G.C. (1992) *The Handbook of Chlorination and Alternative Disinfectants.* 3rd Ed., Van Nostrand Reinhold, New York, N.Y.

Chapter 2
Regulatory and Public Health Concerns

5   Introduction
7   Public Health Concerns
9   Regulatory Concerns
9   National
10   State
12   Canadian Regulatory Concerns

12   Introduction
14   Regulatory Framework
14   Regulations and Guidelines
15   Disinfection Requirements
17 References

INTRODUCTION

Currently, regulators are using many methods to monitor and control the disinfection processes at municipal wastewater treatment plants (WWTPs) and maintain adherence to design standards. Generally, standards are directed toward control of chlorine-based disinfection systems and may include limits on detention time (which vary from 2 hours at average dry-weather flow to 20 minutes at peak flow), mixing requirements, dosage or residual requirements, and even turbidity and upstream process specifications requiring a high degree of disinfection (WEF, 1992). In addition, compliance with specific water quality discharge or receiving water limitations is commonly required in the WWTP's National Pollutant Discharge Elimination System permit. Such limitations normally consist of coliform limitations such as 200 most probable number (MPN) fecal coliform (FC)/100 mL, 240 MPN total coliform (TC)/100 mL, or 2.2 MPN TC/100 mL. As can be seen, these limitations can vary depending on the different state regulations and on the characteristics of the

---

particular receiving water body (for example, designated uses or dilution capacity). Such limitations typically include a geometric mean, or median value, and a maximum value. Also, limitations on effluent chlorine residuals and toxicity are becoming more common in many areas of the U.S.

To determine design criteria and monitoring requirements for a particular area, consultation with appropriate regulatory agencies is necessary. The requirements may apply to average or peak flows either at present or in the foreseeable future. Even if effluent monitoring requirements do not presently exist for a particular area, future requirements may be implemented. For control of disinfection processes, many agencies are increasing their use of effluent coliform monitoring; some states have moved toward *Escherichia coli (E. coli)* as a bacterial indicator for primary contact waters (WEF, 1992).

The importance of considering design standards that are imposed by the regulatory agency cannot be overemphasized. Carefully evaluate manufacturers' claims of inexpensive, nonhazardous, low-maintenance systems in terms of achievable disinfection levels under actual design conditions. Determine whether the disinfection process will be effective for anticipated water quality and operational conditions. Consider factors such as simplicity of operation, degree of operator skills required, frequency of visits, and WWTP locations. Also consider the variation in water quality parameters such as temperature, pH, suspended solids, ammonia-nitrogen and organic nitrogen concentrations, and industrial contributions. Many processes that satisfactorily achieve 200 MPN FC/100 mL (roughly equivalent to 1 000 MPN TC/100 mL) may be inappropriate for highly restrictive requirements. Other processes that are well suited to tertiary effluent produce poor results on unfiltered secondary effluent. Designs that perform well under average water quality conditions may fail when water quality parameters reach the edge of their normal ranges.

When evaluating discharge requirements and design standards, consider effluent toxicity. Chlorine disinfection, if practiced without dechlorination, typically produces a high level of acute toxicity in fish exposed to the effluent. If the dilution to mitigate this effect is inadequate, or effluent standards require zero or almost zero chlorine residual at the point of discharge, effluent dechlorination or use of an alternative method, such as ozone or ultraviolet disinfection, is required.

The effects of chlorination on organic materials in effluents are beginning to receive more attention. The carcinogenic effects of organochlorine and trihalomethane compounds in drinking water are well documented. To date, few regulatory authorities have established effluent limitations for these compounds. If an effluent requiring disinfection eventually enters a potable water system as a raw water supply, particularly with little or no dilution, the design may have to be optimized to reduce required chlorine dosages, or use of alternative disinfectants may have to be considered. Only limited data are available on actual organochlorine and trihalomethane production resulting from the chlorination of wastewater effluents, particularly in nonnitrified effluents.

If it is determined that actual production is significant, compare the advantages of reduced concentrations of halogenated organic compounds achieved through using alternative disinfectants to the monetary cost of the alternative methods. Other important criteria include reliability and effectiveness. Risks to public health from a less-than-optimal disinfection system would probably be greater than risks associated with higher levels of halogenated organic compounds in the effluent. Where this issue will lead is uncertain at this time (WEF, 1992).

PUBLIC HEALTH CONCERNS. Disinfection is practiced using many different means to improve water quality for subsequent downstream use. A stream or other body of water receiving inadequately disinfected wastewater effluent may be contaminated by pathogenic (disease-causing) organisms. Using this receiving water body as a source for the public water supply or for swimming, growing shellfish, or irrigating crops presents an avenue for the propagation and transmission of disease. Unless contamination of the receiving water is prevented by adequate disinfection of wastewater effluents, waterborne diseases may be spread by any one of these uses.

There are millions of coliform bacteria and large numbers of fecal streptococci in a litre of raw domestic wastewater. The numbers of these indicator organisms may vary widely in different wastewaters and, from time to time, in any particular wastewater, as outlined in Table 2.1.

Table 2.1   Typical wastewater influent concentration ranges for pathogenic and indicator organisms (Casson *et al.*, 1990; Rose, 1988; and U.S. EPA, 1979b)

Organism	Minimum, no./100 mL	Maximum, no./100 mL
Total coliforms	1 000 000	—
Fecal coliforms	340 000	49 000 000
Fecal streptococci	64 000	4 500 000
Virus	0.5	10 000
Cryptosporidium oocysts	85	1 370
Giardia cysts	80	320

The organisms from wastewater-contaminated environments of greatest concern to humans are the enteric bacteria, viruses, and the intestinal parasites. Diseases that are spread via water consumption and/or contact can be severe and sometimes crippling. Bacterial diseases such as salmonellosis (including typhoid and paratyphoid fever), cholera, gastroenteritis from enteropathogenic *E. coli,* shigellosis (bacillary dysentery), and viral diseases caused by infectious hepatitis virus, poliovirus, Coxsackie viruses A and B, echoviruses, reoviruses, and adenoviruses may be contracted by contact with, or consumption of, wastewater-contaminated water supplies (that is, potable and/or recreational). Table 2.2 summarizes these diseases and their respective transmission routes.

Table 2.2   Waterborne diseases and transmission routes (Pipes, 1982, and Sorvillo *et al.*, 1992)

Waterborne diseases	Reported in U.S. as transmitted via		
	Drinking water	Recreational waters	Shellfish
Bacterial diseases			
Bacillary dysentery (*Shigella* spp.)	Yes	Yes	No
Cholera (*Vibrio cholerae*)	No	No	Yes
Diarrhea (enteropathogenic *Escherichia coli*)	Yes	No	No
Leptospirosis (*Leptospira* spp.)	Yes	Yes	No
Salmonellosis (*Salmonella* spp.)	Yes	Yes	Yes
Typhoid fever (*Salmonella typhosa*)	Yes	Yes	No
Tularemia (*Francisella tularensis*)	Yes	No	No
Yersinosis (*Yersinia pseudotuberculosis*)	Yes	No	No
Unknown etiology			
Diarrhea, acute undifferentiated	Yes	Yes	Yes
Gastroenteritis, acute, benign, self-limiting[a]	No	Yes	No
Viral diseases			
Gastroenteritis (Norwalk-type agents)	Yes	No	No
Hepatitis A (hepatitis virus)	Yes	Yes	Yes
Parasitic diseases			
Amoebic dysentery (*Entamoeba histolytica*)	Yes	No	No
Ascarariosis (*Ascaris lumbricoides*)	No	No	No
Balantidial dysentery (*Balantidium coli*)	No	No	No
Giardiasis (*Giardia lamblia*)	Yes	No	No
Cryptosporidiosis (Enteric coccidian)	Yes	Yes	No

[a] Not a reportable disease, but transmission via recreational waters has been demonstrated by an epidemiological investigation.

These organisms are accompanied by large numbers of *Pseudomonas* species and other microorganisms such as fungi, including forms that may cause plant, animal, or human diseases, as well as purely saprophytic organisms that are not harmful to health. Many of these latter organisms are beneficial or essential to the success of biological wastewater treatment processes. Typical microorganism ranges for secondary effluent are shown in Table 2.3.

Table 2.3   Secondary effluent ranges for pathogenic and indicator organisms before disinfection (U.S. EPA, 1986)

Organism	Minimum, no./100 mL	Maximum, no./100 mL
Total coliforms	45 000	2 020 000
Fecal coliforms	11 000	1 580 000
Fecal streptococci[a]	2 000	146 000
Viruses	0.05	1 000
Salmonella sp.	12	570

[a]   Assuming removal efficiencies for fecal streptococci similar to the fecal coliform removal efficiencies.

A comparison of Tables 2.1 and 2.3 demonstrates that conventional treatment of domestic wastewater without disinfection cannot be considered sufficient for removal and control of pathogens where public use and body contact occur.

REGULATORY CONCERNS

NATIONAL. In 1972, Congress enacted the most comprehensive water pollution control legislation in U.S. history. The stated objective of the Federal Water Pollution Control Act Amendments of 1972 (Public Law 92-500), as amended in 1977 and 1987 (Clean Water Act, Public Laws 95-217 and 97-117, respectively), was "to restore and maintain the chemical, physical and biological integrity of the Nation's waters." To achieve these objectives, Section 301 of the act mandated the implementation of technology-based effluent standards for industrial and municipal wastewater discharges.

Where higher levels of protection are required, site-specific water quality standards may be used to upgrade effluent limitations. In accordance with these technology-based standards, secondary treatment is the minimum level required for WWTPs.

The level of effluent contaminant reduction achievable by secondary wastewater treatment was defined by the U.S. Environmental Protection Agency (U.S. EPA) in 1973 as Part 133, Title 40, of the Code of Federal Regulations (40 CFR 133), the Secondary Treatment Information regulation. This regulation defined secondary treatment in terms of biochemical oxygen demand, suspended solids, pH, and fecal coliform bacteria. The fecal coliform limitation was deleted in 1976 and replaced by the provision that individual states would develop site-specific water quality standards mandating the disinfection requirements for municipal WWTPs. Much of the concern surrounding disinfection as a part of the secondary treatment definition centered on the undesirable consequences of chlorination practices, namely, the toxic effects of chlorine and its reaction byproducts on freshwater, estuarine, and marine organisms, and the potential formation of carcinogens (U.S. EPA, 1986).

STATE. The responsibility for establishing water quality standards rests with the appropriate state agencies. However, guidance is provided by U.S. EPA and other federal agencies. The first federal water quality criteria for bacteria were proposed in 1968 by the National Technical Advisory Committee (NTAC) of the U.S. Department of the Interior. The criteria for fresh and marine waters were based on epidemiological studies of freshwater quality and health conducted by the U.S. Public Health Service (U.S. PHS). The studies showed that people who swam in water with a median total coliform density of 2 300 TC/100 mL had a significantly greater illness rate than the total control population, which swam in waters with a median total coliform density of less than 1 200 TC/100 mL (U.S. EPA, 1979a and 1984).

The NTAC proposed the use of the fecal coliform indicator because it was indicative of human fecal contamination (therefore, health risk) and less subject to variation. The U.S. PHS studies also showed that approximately 18% of the total coliforms were fecal coliforms. This proportion was used to determine that the equivalent of 2 300 TC/100 mL (the density at which a statistically significant swimming-associated gastrointestinal illness was observed) was approximately 400 MPN FC/100 mL. The NTAC suggested that a detectable risk was undesirable; therefore, a threshold at 50% of the density at which a public health risk occurred was proposed. The fecal coliform criteria based on this threshold include the following (U.S. EPA, 1992):

- Minimum of five samples over a 30-day period,
- Fecal coliform content of primary contact recreation waters not to exceed a geometric mean of 200 MPN/100 mL, and
- Ten percent of the total samples during any 30-day period not to exceed 400 MPN/100 mL.

In 1976, U.S. EPA recommended using the above fecal coliform criteria for establishing microbiological water quality standards. To date, most of the states in the U.S. use this guideline.

However, criticism of the original U.S. PHS studies suggested that U.S. EPA needed to reevaluate this issue. With new research, U.S. EPA hoped to determine whether swimming in wastewater-contaminated waters posed a human health risk, the types of illnesses caused, which bacterial indicators were best correlated with swimming-associated health problems, and the best criteria for determining the health risk. These U.S. EPA studies were conducted in both marine and freshwater areas designated as recreational swimming areas and having well-defined sources of human fecal pollution.

The results of 1972 U.S. EPA studies were based on the strength of the relationship between the rate of gastroenteritis illness caused by viruses in human fecal wastes and the bacterial indicator density as measured with the Pearson correlation coefficient. The Pearson correlation coefficient is a straightforward statistical indicator of the strength of the relationship between two variables, x_1 and x_2, and does not take into account other variables that may be having an effect. The two variables are used to calculate a regression coefficient, y-intercept, and 95% confidence intervals for the paired data. Enterococci were shown to have a strong correlation with swimming-associated gastroenteritis in both marine and fresh waters. *E. coli* showed a strong relationship with gastroenteritis in fresh water but did not show a strong correlation in marine waters. Fecal coliforms and total coliforms showed weak correlations in marine and fresh waters (U.S. EPA, 1992).

The results of 1972 U.S. EPA studies are reflected in the water quality standards for bacteria in a few states that currently use both *E. coli* and enterococci as indicators. Maine and New Hampshire use *E. coli* as their freshwater indicator and enterococcus as their marine indicator. Vermont, Indiana, and Ohio, which do not have marine environments, use *E. coli* only as the bacterial indicator for their freshwater systems (U.S. EPA, 1992).

Some states have adopted even more stringent disinfection requirements than 200 MPN/100 mL for mean effluent fecal coliform levels. Since 1976, the Maryland Department of Health and Mental Hygiene (in the Maryland Water Quality Standards Policy of 1978) has required all facility plans to provide for disinfection to total coliform levels of 3 MPN/100 mL unless a special exemption is granted by the department. This high-level disinfection is applicable to both coastal and inland waters but is geared toward protection of the state's intensive shellfish industry. The California State Department of Health's "Uniform Guidelines for Sewage Disinfection" incorporates dilution, receiving water quality, and beneficial use considerations in the disinfection requirements, which results in varying coliform standards for different discharge situations (Crook, 1984). The standard for nonrestricted recreational uses of wastewater and shallow ocean discharges in close proximity to shellfish areas specifies a 7-day median total coliform value of 2.2 MPN/100 mL or less at some point in the treatment process. The 2.2 MPN TC/100 mL standard applies to those discharge situations in which reclaimed water (that is, 100% wastewater effluent) is impounded for body contact recreational activities or discharged to ephemeral streams with little or no dilution, or where body contact recreation has been designated as a beneficial use of the stream (U.S. EPA, 1986).

In summary, the various states have fecal coliform standards ranging from less than 2.2 MPN/100 mL to as high as 5 000 MPN/100 mL and total coliform standards from less than 2.2 MPN/100 mL to 10 000 MPN/100 mL. Eighteen states (Maine, Vermont, Pennsylvania, Alabama, Georgia, Kentucky, Indiana, Minnesota, Ohio, New Mexico, Oklahoma, Iowa, Missouri, Nebraska, North Dakota, South Dakota, Wyoming, and Idaho) have seasonal disinfection requirements that apply only during the swimming season. These disinfection criteria have been established relative to discharge stream water quality, with the most common standards being fecal coliform limits of 200 MPN/100 mL (full body contact) to 1 000 MPN/100 mL (secondary). More than 40 states have multilevel standards for disinfection relative to the discharge stream water quality criteria, with 200 MPN FC/100 mL being the most common standard. Only four states (Connecticut, Rhode Island, New York, and Utah) have both fecal and total coliform disinfection criteria. Five states (Maine, New Hampshire, Vermont, Indiana, and Ohio) have adopted an *E. coli* standard for fresh water.

This discussion is general in nature and is intended to provide an overview of the general requirements. Site-specific information can be obtained from the specific U.S. EPA regional offices and/or state regulatory offices.

*C*ANADIAN *REGULATORY CONCERNS*

INTRODUCTION. The overall objective of wastewater disinfection in Canada is similar to that in the U.S.—balancing public health protection and environmental risk. In this section, an overview of Canadian requirements for the disinfection of municipal wastewater is provided. The discussion is general in nature and is intended to provide a basic understanding of the requirements in the provinces and territories of Canada. Readers needing detailed and up-to-date information should contact the appropriate regulatory agencies listed in Table 2.4.

Table 2.4 Canadian legislative and regulatory documents and agencies (U.S. EPA, 1979b)

Province or territory	Legislation	Guidelines or regulations	Provincial or territorial regulatory agency
Alberta	Environmental and Enhancement Act	Standards and Guidelines for Municipal Waterworks, Wastewater and Storm Drainage Systems	Alberta Environmental Protection (403) 427-5883
British Columbia	Waste Management Act	Municipal Sewage Discharge Criteria (draft)	Ministry of Environment, Lands and Parks (604) 387-9974
Manitoba	Environmental Act	Surface Water Quality objectives	Manitoba Environment (204) 945-7012
New Brunswick	Clean Environment Act	Guidelines for the Collection and Treatment of Wastewater	Department of the Environment (506) 457-7805
Newfoundland	Department of Environment and Lands Act	Environmental Control Regulations	Department of Environment and Lands (709) 729-1930
Northwest Territory	Waters Act	Guidelines for the Discharge of Treated Municipal Wastewater	Water Resources Division (403) 920-8247
Nova Scotia	Public Health Act Environmental Protection Act	Standards and Guidelines for the Collection, Treatment and Disposal of Sanitary Sewage	Department of the Environment (902) 424-5300

Table 2.4 Canadian legislative and regulatory documents and agencies (U.S. EPA, 1979b) (continued)

Province or territory	Legislation	Guidelines or regulations	Provincial or territorial regulatory agency
Ontario	Water Resources Act	Effluent Disinfection Requirements for Sewage Works Discharging to Surface Waters (draft)	Ministry of Environment (416) 314-3892
Prince Edward Island	Environmental Protection Act	None	Department of the Environment (902) 368-5038
Quebec	Environment Quality Act	Determination of Environmental Objectives for Conventional Parameters for Liquid Effluents	Ministry of the Environment (514) 873-9988
Saskatchewan	Water Pollution Control Act	Surface Water Quality Objectives	Environment and Public Safety (306) 787-6137
Yukon Territory	Waters Act	Waters Regulation	Water Resources Division (403) 667-3100

REGULATORY FRAMEWORK. In Canada, two levels of government have the major legislative and regulatory authority over the environment as well as important leadership roles in integrating the environment and economy. The federal agency is Environment Canada. In addition, each of the 10 provinces has a Ministry of Environment, and the two territories, Yukon and the Northwest Territories, both have Water Resources Divisions. The current broad mandate of the Federal–Provincial–Territorial partnership is "to manage, protect and enhance the quality of the environment for the benefit of present and future generations."

REGULATIONS AND GUIDELINES. Federal legislative documents do not contain specific effluent disinfection requirements. However, the Federal Fisheries Act requires that all effluents discharged directly or indirectly to fish-bearing waters contain no deleterious materials. This implies that the disinfection practice in use should not cause toxic conditions to the native fish population.

Disinfection and monitoring requirements are specified in the various provincial/territorial guidelines or regulations. These documents are prepared under the provisions outlined in the respective legislative acts indicated in Table 2.4.

In addition, some provinces have design standards that specify minimum requirements for handling and storing chlorine and sulfur dioxide containers. These standards may also stipulate dosage and chlorine residual requirements to achieve adequate disinfection.

DISINFECTION REQUIREMENTS. In Canada, the need for disinfection is considered on a case-by-case basis. Generally, bacteriological quality criteria, expressed as total and/or fecal coliform count, are used for determining the need for disinfection. The allowable levels of indicator organisms in receiving waters are dependent on the use of the waters. Some examples are provided in Table 2.5.

Table 2.5   Maximum coliform limits in Canada

Province	Total coliform, MPN/L	Fecal coliform, MPN/L
British Columbia		
Shellfish areas		14 (Proposed)
Recreational areas		200 (Proposed)
Manitoba	1 500	200
New Brunswick		
Water supply	1 000 (At intake)	100 (At intake)
Swimming areas		200
Shellfish areas		14 (Seasonal limit)
Populated areas		200
Newfoundland	5 000	1 000
Ontario		200
Quebec	2 400	400
Saskatchewan		
Noncontact	5 000	1 000
Contact	2 500	200

Where chlorine is used, the design criteria often call for sampling and measurement of total residual chlorine (TRC) concentrations. In Alberta, a TRC concentration of 2.0 mg/L must be maintained after 20 minutes of contact time at peak flow. In the provinces of Nova Scotia and New Brunswick, a TRC concentration of 0.5 mg/L is typically required after 30 minutes of detention time at average flow. Saskatchewan requires the TRC concentration to be between 0.5 and 2.5 mg/L.

Some provinces consider the self-purification process in the receiving stream a possible basis for waiving the need for disinfection. In Ontario, if adequate dilution of the effluent and a long time interval before reuse are present, disinfection may not be necessary.

In lagoons, natural die-off can result in significant coliform reductions. As a general rule, effluent from a properly designed and operated lagoon need not be disinfected.

Full (year-round) or seasonal relaxation of disinfection may be allowed based on a case-by-case evaluation. An environmental impact report may be required to support a request for relaxation. Such reports must include background or predischarge data on the receiving environment. A monitoring program may have to be implemented if adequate background information is not available. The proponent bears the responsibility of demonstrating that disinfection relaxation will not cause adverse effects on the receiving water environment. Generally, relaxation is considered if one or more of the following conditions apply:

- The treatment process includes unit operations with an established capability to consistently achieve the required bacteriological quality without a terminal disinfection stage.
- The effluent is discharged to receiving water not used for recreational, agricultural, or consumptive purposes.
- No adverse effects will result on waters used for recreational, agricultural, or consumptive purposes.

Most large-population centers are incapable of satisfying the above conditions and are required to provide year-round disinfection. Where seasonal relaxation is permitted, year-round disinfection capability is often required in case it becomes necessary.

The provinces of Saskatchewan, Quebec, and Prince Edward Island do not favor the use of chlorine for disinfection. In British Columbia, if chlorine is used, dechlorination is mandatory to eliminate toxic effects on fish. In other provinces and territories, the need for dechlorination is evaluated on a case-by-case basis. In Newfoundland, dechlorination may be required to achieve a maximum allowable TRC of 1 mg/L in the effluent.

REFERENCES

Casson, L.W., *et al.* (1990) *Giardia* in Wastewater. Effect of Treatment. *J. Water Pollut. Control Fed., 62,* 670.

Crook, J. (1984) Wastewater Disinfection in California. In Wastewater Disinfection—The Pros and Cons. Paper presented at 65th Annu. Conf. Water Pollut. Control Fed., Preconference Workshop, New Orleans, La.

Pipes, W.O. (1982) *Bacterial Indicators of Pollution.* CRC Press Inc., Boca Raton, Fla.

Rose, J.B. (1988) Occurrence and Significance of Cryptosporidium in Water. *J. Am. Water Works Assoc., 80,* 2.

Sorvillo, F.J., *et al.* (1992) Swimming-Associated Cryptosporidiosis.

U.S. Environmental Protection Agency (1979a) *Health Effects Criteria for Fresh Recreational Waters.* EPA-600/1-84-004, Cincinnati, Ohio.

U.S. Environmental Protection Agency (1979b) *Health Risks Associated with Wastewater Treatment and Disposal Systems.* EPA-600/1-79-016a, Cincinnati, Ohio.

U.S. Environmental Protection Agency (1984) *Ambient Water Quality Criteria for Bacteria.* EPA-44D/5-84-002, Cincinnati, Ohio.

U.S. Environmental Protection Agency (1986) *Design Manual: Municipal Wastewater Disinfection.* EPA-625/1-86-021, Office Res. Develop., Water Eng. Res. Lab., Cent. Environ. Res. Info., Cincinnati, Ohio.

U.S. Environmental Protection Agency (1992) Draft Report, March 1992. Office Policy Anal., Office of Policy, Planning and Evaluation, Washington, D.C.

Water Environment Federation (1992) *Design of Municipal Wastewater Treatment Plants.* Manual of Practice No. 8, Alexandria, Va.; Am. Soc. Civ. Eng., Report on Engineering Practice No. 76, New York, N.Y.

Chapter 3
Microbial
Fundamentals

20 The Need for Wastewater
        Disinfection
20 Introduction
20 Water Quality Standards and
        Disinfection Criteria
21 History of Waterborne Disease
        Outbreaks
21   Drinking Water
26   Recreational Water
27   Shellfish
27   The Mills–Reincke Phenomenon
        and the Hazen Theorem
28 Need for Disinfection—Summary
28 Pathogens, Diseases, and Microbial
        Fundamentals
28 Common Epidemiological Terms
29 Spread of Infections and
        Preventive Measures
29   Spread of Infections
30   Preventive Measures
30   Immunization
30 Principal Pathogens of Concern
30   Basic Types of Pathogenic
        Organisms
31   Bacteria
31   Fungi
31   Protozoa

31   Algae
32   Viruses
32   Pathogens and Disease
37   Pathogens and Indicator
            Organisms
39   Survival of Pathogens in the
            Environment
40 Wastewater Disinfection
        Technologies
40   Types of Disinfection Technologies
45   Disinfection Technologies and
            Mechanisms of Microbial
            Inactivation
45   Chlorine
45   Ozone
46   Ultraviolet Radiation
47   Microbial Inactivation Unrelated
            to Disinfection
48   Disinfection Kinetics
48   Dose, Residual, Demand, and
                Contact Time
48   Kinetic Models of Disinfection
51   Regrowth Phenomenon
51   Plant Control of Disinfection
52   Residual Toxicity of Disinfectants
54 References

THE NEED FOR WASTEWATER DISINFECTION

INTRODUCTION. Protection of public health through control of disease-causing (pathogenic) microorganisms is the primary reason for wastewater disinfection. This concern probably outweighs other considerations in wastewater treatment (such as reduction of organic matter, inorganics, nutrients, odor, aesthetics, and maintaining the waste-assimilative capacity of receiving water bodies). The relative importance of disinfection in wastewater treatment can be understood by considering that this process is the last barrier protecting receiving waters from pathogenic organisms (U.S. EPA, 1986, and WPCF, 1984). As the population increases and, consequently, greater demands are placed on the water supply, the use of reclaimed water for such things as recreation significantly increases the probability of human exposure to discharged wastewater. Though the concentrations of pathogens decrease naturally with dilution or time, the large volumes of wastewater being discharged and multiplicity of discharge locations typically render the natural means of pathogen reduction insufficient for protecting public health. Hence, wastewater disinfection is necessary to reduce transmission of infectious diseases and safeguard public health.

Disinfection of wastewater is practiced to protect water quality for subsequent downstream use. A body of water receiving inadequately disinfected wastewater may be contaminated with pathogenic organisms. Diseases may result from the use of such water as a source of public water supply or for bathing, producing shellfish, or irrigating crops.

WATER QUALITY STANDARDS AND DISINFECTION CRITERIA.
The process for setting water quality guidelines and standards has historically followed a pattern of establishing the effects of contaminants on human health and the ecology of aquatic organisms (Cabelli, 1980). The development of these guidelines and standards, the first step in the process, has largely been dictated by the level of water quality attainable using the Best Available Technology. Guidelines have been based, in many cases, on limited epidemiological and ecological evidence using little, if any, data to quantify risk in relation to pollutant levels in the environment. The second step has been the modification of the guidelines and standards on the basis of detectable risk, using a limited quantity of data. The last step has been the development of guidelines based on acceptable risk, which requires an epidemiological or ecological database broad enough to model mathematically the relationship between some measure of water quality and the risk (U.S. EPA, 1986).

Various water quality standards and disinfection criteria have been established on both the federal and state levels. Refer to Chapter 2 for further details on the regulatory aspects of effluent disinfection.

HISTORY OF WATERBORNE DISEASE OUTBREAKS. Some diseases are epidemic, affecting many people in a community simultaneously. Examples of epidemic diseases are typhoid fever, cholera, dysentery, and gastroenteritis. Epidemics may result from contamination of water supplies by wastewater. Evidence of epidemic disease has diminished with the wider use of wastewater treatment, improvements in water purification, and other sanitary measures.

The Centers for Disease Control, U.S. Environmental Protection Agency, and state and local health agencies share data on waterborne diseases. Data on the number of reported outbreaks and cases of waterborne diseases in the U.S. for the period 1971–1980 are summarized in Table 3.1.

These data apply only to those outbreaks that have been reported to health agencies; the true incidence is probably greater than that reported. Also, municipal wastewater represents only one of many sources of pathogens related to disposal of human wastes. Other sources include, but are not limited to, septic tank discharges, leach field failures, personal contact, and cross-connections.

Drinking Water. Data for disease outbreaks (related directly to drinking water) during 1980 and 1981 are presented in Table 3.2. In 1980, 50 outbreaks, with a total of 20 008 cases, were reported. Acute gastrointestinal illness of undefined etiology was identified in 28 outbreaks, with 13 220 cases.

In 1981, fewer outbreaks were reported (4 430 cases) compared to 1980. Acute gastrointestinal illness was still the most common disease. One cholera outbreak of 17 cases was reported, the largest cholera outbreak in the U.S. in this century (WPCF, 1984).

Most outbreaks have been associated with noncommunity water supplies. In general, the number of cases in each of these outbreaks is low. However, outbreaks associated with community systems, though relatively rare, typically affect many more people. For example, the outbreak of cryptosporidiosis in Milwaukee, Wisconsin (March and April of 1993), resulted in cases of watery diarrhea in approximately 403 000 people (Moore *et al.*, 1994).

Waterborne disease outbreaks are associated with heavily contaminated water that receives little or no treatment in noncommunity systems or deficient or poor treatment in community systems. Deficiencies in water treatment before consumption have been by far the most common reasons for outbreaks (18 out of 50 outbreaks in 1980, 11 out of 32 in 1981, and 17 out of 50 in 1991 to 1992). While untreated wastewater or runoff containing raw wastewater has caused many outbreaks, the discharge of treated wastewater has rarely resulted in disease outbreaks associated with drinking water in the U.S.

Table 3.1   Waterborne disease outbreaks, 1971 to 1980, by type of illness and water system

Disease	Community systems		Noncommunity systems		Individual systems		Total	
	Outbreaks	Cases	Outbreaks	Cases	Outbreaks	Cases	Outbreaks	Cases
Gastroenteritis, unidentified etiology	55	28 928	110	10 783	12	134	177	39 845
Giardiasis	22	17 090	12	2 390	5	72	39	19 552
Chemical poisoning	22	2 886	7	645	9	63	38	3 594
Shigellosis	10	3 788	12	1 401	2	6	24	5 195
Salmonellosis	5	1 075	2	72	1	3	8	1 150
Hepatitis A	5	130	7	305	4	28	16	463
Campylobacter diarrhea	2	3 800	—	—	1	21	3	3 821
Viral gastroenteritis	2	1 690	8	1 457	—	—	10	3 147
Typhoid	—	—	1	210	3	12	4	6 222
Toxigenic *E. coli* diarrhea	—	—	1	1 000	—	—	1	1 000
Total	123	59 387	160	18 263	37	339	320	77 989

Table 3.2 Waterborne disease outbreaks related to drinking water during 1980 and 1981 (WPCF, 1984)

Disease or agent	1980 Outbreaks	1980 Cases	1981 Outbreaks	1981 Cases
Acute gastrointestinal illness	28	13 220	14	1 893
Giardia	7	1 724	9	297
Chemical	7	2 298	5	128
Shigella	1	4	1	253
Campylobacter	1	800	1	81
V. cholerae 01	nr[a]	nr	1	17
Rotavirus	nr	nr	1	1 761
Norwalk agent	5	1 914	nr	nr
Hepatitis	1	48	nr	nr
Total	50	20 008	32	4 430

[a] nr = none reported.

(WPCF, 1984). One of the reasons for this has been terminal disinfection practiced at wastewater treatment plants (WWTPs) (WPCF, 1984).

Where wastewater disinfection is not practiced, disease can be transmitted to humans by circuitous routes. An outbreak of *Salmonella paratyphi* B on a dairy farm in Yorkshire, England, was traced to treated wastewater from a village where a human carrier lived. *Salmonella paratyphi* B was isolated from both the raw and treated wastewater. The effluent was discharged into a cow pasture, the dairy herd became infected, and the infection was transmitted to humans on the farm. *Salmonella typhimurium*, *Salmonella derby,* and *Salmonella bredency* were also isolated from dairy herds in the area and traced to treated wastewater (WPCF, 1984).

In 1991 and 1992, 17 states and territories reported 34 outbreaks of waterborne disease affecting 17 464 people (Moore *et al.,* 1994). The infectious agent was unknown in 23 of the cases. In seven cases, disease was linked to the protozoan parasites *Giardia* or *Cryptosporidium*. Table 3.3 contains data concerning outbreaks of disease associated with drinking water during the period 1991 to 1992.

Overall, the number of waterborne disease outbreaks has decreased during the period 1971 to 1992 (Figure 3.1), with the majority linked to noncommunity water systems. Over this same period, outbreaks linked to protozoa have become increasingly important (Moore *et al.,* 1994). *Giardia* and *Cryptosporidium* are found in high concentrations in wastewater effluents and may also be associated with nonpoint source contamination (LeChevallier *et al.,* 1991, and Rose *et al.,* 1989).

Table 3.3   Waterborne disease outbreaks related to drinking water during 1991 and 1992 (Moore et al., 1994)

Disease or agent	Community systems		Noncommunity systems		Individual systems		Total	
	Outbreaks	Cases	Outbreaks	Cases	Outbreaks	Cases	Outbreaks	Cases
Gastroenteritis, unidentified etiology	3	10 077	19	3 252	1	38	23	13 367
Giardiasis	2	95	2	28	0	0	4	123
Cryptosporidiosis	2	3 000	1	551	0	0	3	3 551
Hepatitis A	0	0	0	0	1	10	1	10
Shigellosis	0	0	1	150	0	0	1	150
Nitrate	0	0	0	0	1	1	1	1
Fluoride	1	262	0	0	0	0	1	262
Total	8	13 434	23	3 981	3	49	34	17 464

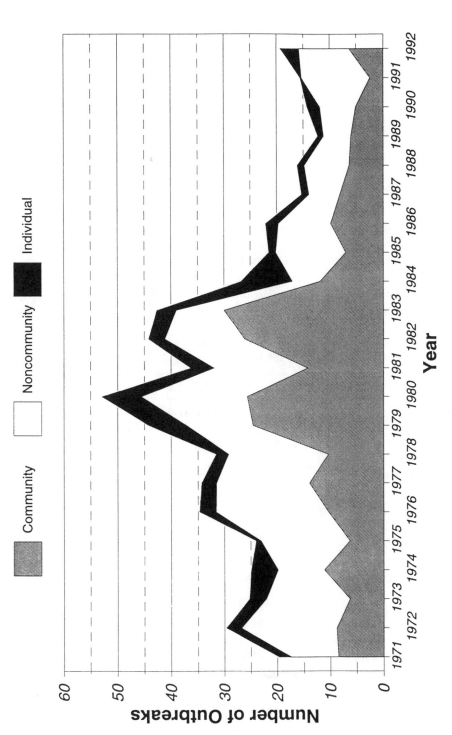

Figure 3.1   Incidence of waterborne disease from 1971 to 1992 as a function of water system (total number of outbreaks = 609) (reprinted from *Journal American Water Works Association*, Vol. 86, No. 2 [February 1994], by permission; copyright ©1994, American Water Works Association)

Recreational Water. Waterborne disease outbreaks in 1980 and 1981 related to recreational activities are shown in Table 3.4. Skin diseases were the most common outbreaks associated with recreational waters. In 1980, however, four outbreaks of shigellosis and one of acute gastrointestinal illness were reported. In 1981, no enteric outbreaks were reported. Infectious hepatitis and salmonellosis have also reportedly been transmitted by recreational contact (WPCF, 1984).

In the U.S., both fresh and marine waters are used extensively for contact recreation. Early studies in the United Kingdom suggested that the risk to health from bathing in wastewater-contaminated waters was negligible (Moore, 1954a and 1954b). The diseases followed in these studies were poliomyelitis and salmonellosis. It is apparent, after almost 50 years of epidemiological study, that poliomyelitis is not waterborne and, thus, would not be expected to be transmitted by contact recreation (WPCF, 1984). Salmonellosis has been transmitted by water, but the number of *Salmonella* bacteria that must be ingested to cause disease (infective dose) is greater than 10^4 (Bryan, 1974, and WPCF, 1984). A bather would not be likely to consume the volume of recreational water containing this number of *Salmonella* bacteria. Thus, it is not surprising that early studies based on these diseases concluded that health risks from bathing were negligible (WPCF, 1984).

Table 3.4   **Disease outbreaks related to recreational water during 1980 and 1981 (WPCF, 1984)**

	1980		1981	
Disease or agent	**Outbreaks**	**Cases**	**Outbreaks**	**Cases**
Dermatitis[a]	5	78[a]	7	642
Shigella	4	335	nr[b]	nr
Acute gastrointestinal illness	2	83	nr	nr
Adenovirus	1	15	nr	nr
Legionella	nr	nr	1	34
Total	12	511[a]	8	676

[a] Four outbreaks of dermatitis were attributed to *Pseudomonas aeruginosa*, and the agent for one outbreak was undetermined.
[b] nr = none reported.

A series of field studies to determine the relationship between health and bathing-water quality in the U.S. demonstrated that swimmers had a higher overall incidence of disease than nonswimmers, regardless of water quality (Stevenson, 1953). Swimmers younger than 10 years of age had almost twice the illness rate of those older than 10. A 99% correlation between illness and bathing-water quality was shown for Lake Michigan in Chicago, Illinois.

However, the evidence was not considered conclusive because of the short duration of the study. A significant increase in gastrointestinal disturbances was observed in swimmers in the Ohio River, where the total coliform level was 2 700/100 mL (WPCF, 1984).

Studies conducted by Cabelli (1977) and Cabelli *et al.* (1975a, 1975b, and 1976) suggest that enteric disease is transmitted by contact recreation and that the level of disease is related to the level of contamination as indicated by the density of enterococci (WPCF, 1984).

Other outbreaks demonstrate that other enteric diseases are transmitted by contact recreation. A 1974 shigellosis outbreak in Dubuque, Iowa, was related to swimming in the Mississippi River (Rosenberg *et al.*, 1975). The mean level of fecal coliforms in the water was 17 500/100 mL, with counts as high as 5×10^6/100 mL downstream from the WWTP and 4×10^5/100 mL at the beach. The disinfection at the WWTP was marginal at best (WPCF, 1984).

Ingestion during swimming resulted in 13 outbreaks of gastroenteritis in 1989–1990 and 11 outbreaks in 1991–1992 (Moore *et al.,* 1994). While the number of outbreaks is similar to that reported in 1980 and 1981, the etiologic agents most commonly associated with the disease have changed. Unlike past years when bacterial and viral agents were solely responsible, 55% of the recently reported outbreaks have been linked to the protozoa *Giardia* and *Cryptosporidium*.

Shellfish. Protecting shellfish-harvesting regions is an important reason for wastewater disinfection. The combination of reductions in pathogenic bacteria, viruses, and parasite populations; stipulation of a no-harvest buffer zone below WWTPs; and implementation of modern shellfish sanitation programs has successfully curtailed the number of disease outbreaks associated with shellfish consumption (WPCF, 1984). A report to Congress by the Comptroller General of the United States (1977), while critical of wastewater disinfection with chlorine, recognized the value of disinfecting treated wastewater for protection of shellfish-growing areas (WPCF, 1984).

The Mills–Reincke Phenomenon and the Hazen Theorem. Mills and Reincke reported in 1893 that when a polluted water supply was replaced by a pure supply, the overall health of the community improved to a greater degree than could be attributed to the reduced prevalence of typhoid fever and other recognized waterborne diseases (White, 1992). This discovery became known as the Mills–Reincke phenomenon. Allen Hazen discovered in 1903 that when a community water supply changed from bad to excellent because of adequate treatment, for every person saved from death by typhoid, three other persons would be saved from death by other causes, many of which were probably never considered to have any connection with, or be especially affected or influenced by, the quality of the public water supply (White, 1992). This decrease in the death rate was known as the Hazen theorem. Clearly, disinfection of public water supplies and wastewater has greater implications than just the control of waterborne diseases (White, 1992).

NEED FOR DISINFECTION—SUMMARY. Disinfecting wastewater reduces the number of disease-causing microorganisms in discharges from WWTPs and minimizes their dissemination in the receiving water. Terminal disinfection minimizes the stress (reduces microbial loads) on downstream water treatment plants. Nevertheless, outbreaks of disease from inadequately treated drinking water are reported each year (WPCF, 1984, and Moore *et al.,* 1994) as a result of nonpoint source pollution and improper operation of water treatment facilities.

While there is still considerable debate over whether enteric disease is transmitted by contact recreation, recent epidemics and epidemiological studies indicate that recreation is, in fact, associated with disease transmission (WPCF, 1984, and Moore *et al.,* 1994).

PATHOGENS, DISEASES, AND MICROBIAL FUNDAMENTALS

COMMON EPIDEMIOLOGICAL TERMS. Several common epidemiological terms are relevant to a discussion of wastewater disinfection and its effect on public health:

- *Disease vectors* are agents that transmit pathogens from one organism to another either mechanically, as a carrier, or biologically, with a role in the life cycle. The housefly, for example, is a mechanical carrier of the disease agents of typhoid fever (*Salmonella typhi*), and mosquitoes can transmit disease agents causing malaria and yellow fever. The occurrence of a disease does not necessarily require vectors: humans can become sick from direct ingestion of, or contact with, contaminated water.
- *Routes* are the transmission pathways of pathogens from their sources to humans—for example, the anal–oral route. The source could be *primary,* such as the feces of infected humans or animals (typically called the *reservoir*), or *secondary,* such as raw or treated wastewater. Vectors and sources are also often referred to as *carriers.*
- *Infective dose* is typically expressed as the number of pathogenic organisms estimated to result in human infection. For example, intake of *Salmonella typhi* will not necessarily cause typhoid in the host; an infective dose of the organism is necessary to do so. Hence, evidence of low numbers of pathogenic organisms in treated wastewater should not alarm the public unless there is a possibility of ingesting the infective dose. Viral infections are believed to have low infective doses

(often only one virus) because viruses often can multiply readily in a host cell or organism.

- *Incubation period* is the time between intake of the infective dose and manifestation of the symptoms of disease in the host. During the incubation period, the pathogens multiply in the body and overcome normal immune system responses. For some infections, the body's immune system may eventually recover to the extent that the disease becomes transitory, with medication being unnecessary.

SPREAD OF INFECTIONS AND PREVENTIVE MEASURES. Spread of Infections. There are three basic routes through which infection can occur: ingestion from splashes, contaminated food, water, or cigarettes; inhalation of infectious agents or aerosols; and infection of an unprotected cut or abrasion. Workers at WWTPs often come in physical contact with raw wastewater and sludge in the course of their daily activities. Even when direct physical contact is avoided, a worker may handle contaminated objects. Cuts and abrasions, however minor, should be properly treated. Open wounds invite infection from many of the viruses and bacteria present in wastewater (WPCF, 1991). Table 3.5 summarizes the major routes of infection.

Table 3.5   Routes of infection (WEF, 1991)

Indirect contact	
Ingestion	Eating, drinking, or accidentally swallowing pathogenic organisms (for example, hepatitis A)
Direct contact	
Inhalation	Breathing spray or mist containing pathogenic organisms (for example, common cold)
Accidental contact	
Break in skin	Entry of pathogenic organism to body via contact, cut, or break in the skin (for example, tetanus)

The *direct contact* route may involve touching or inhaling airborne microbes (U.S. EPA, 1992). The *indirect contact* route may involve the following possibilities (U.S. EPA, 1992):

- Consumption of pathogen-contaminated crops or food products;
- Ingestion of contaminated drinking water or recreational waters;
- Consumption of inadequately cooked or uncooked pathogen-contaminated fish or shellfish; and
- Contact with wastewater pathogens transported by rodents, insects, or other vectors, including grazing animals.

Ingestion is generally the major route of infection. Simply touching the mouth with a hand contributes to the possibility of infection. Workers who eat or smoke without washing their hands have a much higher risk of infection. Most surfaces near wastewater equipment are likely to be covered with bacteria or viruses. These potentially infectious agents may be deposited on surfaces in the form of an aerosol or may come from direct contact with the wastewater or sludge (WPCF, 1991).

Prolonged exposure to aerosols should be avoided. Use of surgical masks and goggles may help minimize contact with aerosols.

Preventive Measures. Knowledge of the transmission routes and means by which diseases can spread helps plant personnel take precautions to significantly reduce disease transmission via wastewater. These precautions include common personal protection measures recommended for all employees at the workplace, safe laboratory practices, and adequate first-aid treatment and immunizations. Among the common personal protection measures are the use of common sense, awareness of hazards, and implementation of safety precautions.

Laboratory personnel should be provided training in proper microbiological techniques and safety. Workers should be familiar with aseptic handling techniques and the biology of organisms. Appropriate vaccinations should be administered to the workers if known pathogens are being evaluated. Laboratory apparatus and waste should be properly decontaminated or sterilized to minimize the risk of disease transfer (WPCF, 1991).

IMMUNIZATION. Immunization is the process of providing vaccination to the public so that they are able to resist (be immune to) infection temporarily or permanently. Immunity acquired by vaccination contrasts with natural immunity such that the body does not need an outside agent to combat disease. In acquired immunity through vaccinations, a supply of adequate antigens or antibodies is developed in the body to counteract specific pathogenic microorganisms. Depending on the situation, immunizations may be provided to the general public or wastewater workers to combat various diseases (for example, diphtheria, tetanus, polio, typhoid, hepatitis A, and hepatitis B).

PRINCIPAL PATHOGENS OF CONCERN. Basic Types of Pathogenic Organisms. Diseases are caused by a multitude of microorganisms that are broadly classified under various categories based on some of their common microbial characteristics. These classifications are bacteria, viruses, protozoa, helminths, algae, and fungi. Not all of these microorganisms necessarily cause diseases. Many are essential to the success of biological wastewater treatment processes. Microorganisms are grouped based on the presence or absence of a defined nucleus. The nucleus is a membrane-enclosed organelle that contains chromosomes—the genetic material of the cell (deoxyribonucleic acid, or

DNA). Organisms having a true membrane-bound nucleus are called *eucaryotes*, and those with chromosomes that are not enclosed by a membrane are called *procaryotes*. Bacteria are procaryotes, while protozoa, algae, and fungi are eucaryotes. Viruses are noncellular and are classified separately (Gaudy and Gaudy, 1980).

BACTERIA. Bacteria are a large class of microscopic unicellular organisms that lack plantlike chlorophyll and a membrane-bound nucleus. These microorganisms are often motile, by means of flagella, and occur in three main morphologies: spherical (coccus), rod-shaped (bacillus), and spiral (spirilla). They reproduce mainly by fission and by asexual spore formation. They are ubiquitous in nature and occur mainly in water, soil, organic matter, or living bodies of plants and animals. They can be saprophytic, parasitic, or autotrophic. Some bacteria are pathogenic.

These procaryotic organisms essentially contain a cell wall, a cytoplasmic membrane, DNA, ribonucleic acid (RNA), ribosomes, and cytoplasm. Bacteria can be divided into two major groups, Gram-positive and Gram-negative, on the basis of the Gram's stain reaction that correlates with the structure of the cell wall (Gaudy and Gaudy, 1980). Bacteria commonly range from less than 0.2 μm in size to as large as 15 μm. Spherical bacteria are typically 0.5 to 1.0 μm in diameter, rod-shaped bacteria are typically 0.5 to 1.0 μm in width by 1.5 to 3.0 μm in length, and spiral-shaped bacteria are 0.5 to 5 μm in width by 6 to 15 μm in length (Metcalf & Eddy, Inc., 1991).

FUNGI. Fungi are eucaryotes. They are considered multicellular, nonphotosynthetic heterotrophs. Fungi reproduce sexually or asexually, by fission, budding, or spore formation. The predominant form of fungi growth is filamentous. Molds, or "true fungi," produce microscopic units (*hyphae*) that collectively form a filamentous mass called the *mycelium.* Yeasts are fungi that cannot form a mycelium and, therefore, are unicellular. Most fungi are aerobic and can grow under low-moisture conditions, can tolerate relatively low pH, and have low nitrogen requirements (Metcalf & Eddy, Inc., 1991).

PROTOZOA. Protozoa are single-celled, microscopic eucaryotes. Many undergo complex life cycles. A majority of protozoa are aerobic heterotrophs, although a few are anaerobic. Some are photoautotrophic. The four categories of protozoa are flagellates, sporozoa, ciliates, and rhizopods. Protozoa are generally larger than bacteria and often consume bacteria as an energy source (Gaudy and Gaudy, 1980, and Metcalf & Eddy, Inc., 1991).

ALGAE. Algae are also eucaryotes. These autotrophic, photosynthetic microorganisms can be either unicellular or multicellular. Many algae contain one or more pigments that mask their chlorophyll, thus producing different colors. Algae provide an important source of food for a variety of animals.

VIRUSES. Viruses are submicroscopic entities consisting principally of nucleoprotein and are able to pass through bacteria-retaining filters. Viruses have many characteristics of living organisms (for example, they are capable of growth and multiplication in living cells). They are recognized by their toxic or pathogenic effects on bacteria, plants, and animals. Viruses differ from cellular organisms in several important ways (Gaudy and Gaudy, 1980):

- They are much simpler than other microorganisms in structure and composition. The two essential components of a virus are proteins and nucleic acids, and many contain no other compounds. Other viruses are composed of proteins and nucleic acids surrounded by a membrane of lipid and protein.

- All cells contain both RNA and DNA, but a virus has only one type of nucleic acid. For viruses that contain RNA, the genetic information is carried by RNA rather than DNA. Cellular DNA is a double-stranded molecule that is circular in bacteria, and cellular RNA is single stranded. Viral nucleic acids occur in several forms. The DNA may be single or double stranded, linear, or circular. Though RNA is almost always single stranded, a few viruses contain double-stranded RNA.

- Viruses have no metabolic enzymes, use no nutrients, and produce no energy. They are obligate intracellular parasites and can reproduce only inside a living cell. The parasitism of viruses is unlike that of parasitic bacteria, protozoa, or fungi, which receive nutrients from the host cell. In a cell infected by a virus, the metabolism of the cell is diverted from the manufacture of new cell material to the manufacture of new viruses. Outside a cell, the virus is completely inert and inactive.

- Viruses do not reproduce, as cells do, by increasing in size and dividing. When a virus infects a cell, the proteins and nucleic acids are separated (that is, the virus no longer exists as an entity, and often only the nucleic acids enter the cell). Under the direction of the viral nucleic acids, the cell produces the viral proteins, the viral nucleic acids are replicated, and new viruses are assembled from the parts.

Pathogens and Disease. The diseases of concern in drinking water, contact recreational water, and shellfish follow the anal–oral route of transmission. The clinical and epidemiological features of these diseases are shown in Table 3.6 (Sobsey and Olsen, 1983, and WPCF, 1984). Bacteria are responsible for diseases ranging from mild gastrointestinal upset to typhoid fever.

Table 3.6   Summary of taxonomic, clinical, and epidemiological features of potential drinking water disease agents (WPCF, 1984)

Name of organism or group	Number of types	Major disease	Primary sources in major reservoirs	Concentration in primary source	Infectious dose	Prevalence (average excretion, %)	Duration of shedding	Carrier state[a]
Bacteria								
Salmonella typhi	1	Typhoid fever	Human feces	10^6/g	Low		?[b]	
Salmonella paratyphi	1	Paratyphoid fever	Human feces	10^6/g	High	0.1	?	+
Other salmonellae	2 000	Salmonellosis	Human/animal feces	10^6/g	High		?	
Shigella	4	Bacillary dysentery	Human feces	10^6/g	Medium			+(?)
Vibrio cholerae	?	Cholera	Human feces	10^6/g	High		?	?
Enteropathogenic *E. coli*	?	Gastroenteritis	Human feces	10^8/g	High	1.2–15.5	?	+
Yersinia enterocolitica	?	Gastroenteritis	Human/animal feces	?	High	?	?	?
Campylobacter jejuni	1 000	Gastroenteritis	Human/animal (?) feces	?	?	?	1–3 weeks	?
Legionella pneumophila and related bacteria	7	Acute respiratory illness (legionellosis)	Thermally enriched waters	?	High (?)	?	?	?
Mycobacterium tuberculosis	1	Tuberculosis	Human respiratory exudates	?	?	?	?	
Other (atypical) mycobacteria	2	Pulmonary illness	Soil and water	?	?	?	?	?
Opportunistic bacteria	?	Variable	Natural waters	?	?	?	?	?
Enteric viruses								
Enteroviruses								
Polioviruses	3	Poliomyelitis	Human feces					
Coxsackieviruses A	23	Aseptic meningitis	Human feces	10^6/g				
Coxsackieviruses B	6	Aseptic meningitis	Human feces			1–5%	1–3 weeks	–
Echovirus	31	Aseptic meningitis	Human feces					
Other enteroviruses	4	AHC; encephalitis	Human feces					

Table 3.6 Summary of taxonomic, clinical, and epidemiological features of potential drinking water disease agents (WPCF, 1984) (continued)

Name of organism or group	Number of types	Major disease	Primary sources in major reservoirs	Concentration in primary source	Infectious dose	Prevalence (average excretion, %)	Duration of shedding	Carrier state[a]
Reoviruses	3	Mild upper respiratory and gastrointestinal illness	Human/animal feces	10^9/g	Low	?	1–2 weeks	–
Rotaviruses	3	Gastroenteritis	Human feces		?	?	1 week	–
Adenoviruses	37	Upper respiratory and gastrointestinal illness	Human feces	10^6/g	?	?	?	?
Hepatitis A virus	1	Infectious hepatitis	Human feces			?	3 weeks	–
Norwalk and related gastrointestinal viruses	3	Gastroenteritis	Human feces			?	1 week (?)	–
Protozoans								
Acanthamoeba castellani	1	Amebic meningoence	Soil and water	?	?	Very low	None	+
Balantidium coli	1	Balantidiasis (dysentery)	Human feces	?	?	Very low	Variable	?
Entamoeba histolytica	1	Amebic dysentery	Human feces	10^5/g	Low	3–10	Variable	+
Giardia lamblia	1	Giardiasis (gastroenteritis)	Human/animal feces	10^5/g	Low	1.5–2.0	6–7 weeks	+ (months–years)
Naegleria fowleri	1	Primary amebic illness	Soil and water	?	?	Very low	None	–
Helminths								
Nematodes (roundworms)								
Ascaris lumbricoides	1	Ascariasis	Human/animal feces	10–10^4/g		1	?	+
Trichuris trichiura	1	Trichuriasis	Human feces	10–10^4/g		1	?	+

Table 3.6   Summary of taxonomic, clinical, and epidemiological features of potential drinking water disease agents (WPCF, 1984) (continued)

Name of organism or group	Number of types	Major disease	Primary sources in major reservoirs	Concentration in primary source	Infectious dose	Prevalence (average excretion, %)	Duration of shedding	Carrier state[a]
Hookworms								
Ancylostoma duodenale	1	Hookworm disease	Human feces					
Necator americanus	1	Hookworm disease	Human feces	$10–10^2$		1		+
Strongyloides stercoralis	1	Threadworm disease	Human/animal feces			1		
Algae (blue-green)								
Microcystis aeruginosa	4	Gastroenteritis	Natural waters	$10^4–10^6$	?	–	–	–
Anabaena flos-aqua	6[c,d]	Gastroenteritis	Natural waters	$10^4–10^6$	?	–	–	–
Aphanizomenon flos-aqua	1[d]	Gastroenteritis	Natural waters	$10^4–10^6$	?	–	–	–
Schizothrix calcicola	1	Gastroenteritis	Natural waters	$10^4–10^6$	?	–	–	–

a   Untreated subclinical and/or asymptomatic infections.
b   ? = unknown or uncertain.
c   Numbers refer to number of different toxic strains.
d   Also produces neurotoxins, whose role in human disease is uncertain.

In general, high numbers of enteric pathogens are shed in feces by a small percentage of the population. The shedding of pathogens typically lasts several weeks, but a carrier state has been observed for members of the genuses *Salmonella* and *Shigella*. The infectious dose of the enteric bacteria for healthy individuals generally exceeds 10 000 cells. However, much lower levels of *Shigella* and *Salmonella typhi* can cause infection (WPCF, 1984).

Enteric viruses are shed in feces at high levels for several weeks. Levels of 10^9/g to 10^{10}/g have been reported (Banatvala, 1981). Many studies have been conducted to determine the exact quantity of enteroviruses that must be ingested to cause infection. Poliovirus infection by the oral route has been studied, and Sabin (1957) reported that nonhuman primates and humans did not have comparable susceptibility (WPCF, 1984). Nevertheless, the potential for infection warrants concern, though infection does not always lead to disease.

Waterborne disease caused by protozoa has been primarily associated with *Entamoeba histolytica, Giardia lamblia,* and *Cryptosporidium parvum.* From 1978 to 1981, *Giardia lamblia,* which causes severe gastrointestinal upset, was the most frequently identified pathogen associated with waterborne disease. In 1981, 28% (9 of 32) of the reported outbreaks were caused by *Giardia*. In 1991 and 1992, *Giardia* accounted for 4 of 34 outbreaks (Moore *et al.,* 1994). Waterborne amoebic dysentery has not been reported in the U.S. for some time (WPCF, 1984).

The cysts of *Giardia lamblia* and *Entamoeba histolytica* have been estimated to occur at 10^5/g in feces. Akin *et al.* (1978) reported levels of *Giardia* as high as 2.2×10^6/g in infected children and 9.7×10^7/g in asymptomatic (without symptoms) carriers. *Giardia* was shed for 6 to 7 weeks; 1 to 2% of the human population appeared to be shedding the cysts (WPCF, 1984).

The protozoan parasite *Cryptosporidium* has only been identified as a major disease-causing agent in the last decade. Before 1980, only 11 cases of cryptosporidiosis had been described worldwide (Curds, 1992). Most recent evidence implicates *Cryptosporidium* as the most common parasite found worldwide in patients with diarrhea (Curds, 1992). In 1991 and 1992, *Cryptosporidium* was linked to three outbreaks of waterborne disease in the U.S. (Moore *et al.,* 1994). However, outbreaks of cryptosporidiosis are probably underrecognized (Moore *et al.,* 1994). Immunologically healthy persons with cryptosporidiosis typically have symptoms for fewer than 20 days, but the cysts of *Cryptosporidium* (called *oocysts*) can be excreted for twice as long as the diarrheal period (Fayer and Ungar, 1986). *Cryptosporidium* oocysts occur in numbers ranging from 10^5/g to 10^7/g in calf feces (an animal host for *C. parvum,* a species that infects humans) (Ongerth and Stibbs, 1987). Rose *et al.* (1991) reported *Cryptosporidium* oocysts in 65% of 188 surface water samples collected in 17 states in the U.S., with an average concentration of 43 oocysts per 100 L. While this appears to be a low concentration, 170 oocysts/100 mL has been reported in wastewater (Rose *et al.,* 1989). Feeding studies have indicated that the infective dose for humans is approximately 130 oocysts, with as few as 30 oocysts causing infection in one of five subjects (DuPont *et al.,* 1995).

In the U.S., the transmission of diseases spread by worms (helminths) in water is rare. Parasitic helminths must pass a number of barriers to cause infection and disease in humans. They must be viable and, in many cases, must have the opportunity to develop to the infective stage of their life cycles. After being released from the host, parasitic helminths may require simple or complex periods of incubation in the environment before being infective to a new host. With few exceptions, one or more intermediate hosts are required in the life cycle of the tapeworms that infect humans (WPCF, 1984).

With the increase in the incidence of acquired immunodeficiency syndrome (AIDS), research has been conducted to determine the environmental fate of the human immunodeficiency virus (HIV) that causes AIDS. The HIV is contained in the body fluids (for example, blood and semen) and excretions of infected persons; thus, it can be present in raw wastewater. Casson *et al.* (1992) reported that HIV inoculated into primary effluent and nonchlorinated secondary effluent was stable for as long as 12 hours in the laboratory, but decreased 2 to 3 logs in infectivity within 48 hours. Riggs (1989) noted that the sensitivity of HIV to heat, drying, and disinfectants such as chlorine made it highly unlikely that waterborne transmission of the disease will occur. Further information regarding HIV in wastewater can be obtained from a recent article by Johnson *et al.* (1994).

A few studies have been conducted to determine whether HIV is inactivated by common disinfectants. Spire *et al.* (1984) found that the inactivation of HIV was similar to that of other enveloped viruses. Sodium hypochlorite is an effective disinfectant for HIV (McDougal *et al.,* 1985). A 0.1% (v/v) solution of household bleach (52 mg NaOCl/L) reduced the concentration of HIV 4 logs in 10 to 20 seconds. However, Spire *et al.* (1985) found that ultraviolet (UV) radiation was not effective at inactivating HIV.

The proceedings of a symposium held in 1991 on the survival of HIV in environmental waters contain methods using gene probes and polymerase chain reaction techniques to detect HIV in wastes (Farzadegan, 1991).

PATHOGENS AND INDICATOR ORGANISMS. Because of the potential presence of multitudes of pathogenic organisms in wastewater, routine monitoring for all types of organisms would be prohibitively expensive. Hence, the presence, absence, or quantity of pathogenic organisms has largely been estimated using indicator organisms such as total coliform, fecal coliform, *E. coli,* or *Streptococcus.* Among these, total coliform and fecal coliform in wastewater treatment have been widely used. Laboratory enumeration of these indicator organisms involves growth of the organisms in a suitable culture medium with optimal temperature and pH conditions. Various test procedures such as membrane filtration and multiple-tube fermentation are available. Details of these techniques are found in *Standard Methods for the Examination of Water and Wastewater* (APHA, 1995).

The levels of pathogenic or indicator organisms in raw wastewater and before disinfection are critical parameters for the design of a wastewater disinfection system. Typical levels of these organisms in the raw influents of WWTPs are presented in Chapter 2 (Table 2.1).

Note that the typical range of concentrations shown in Table 2.1 is several orders of magnitude. The possibility of such a wide variation is also supported through observations of other researchers (Geldreich, 1966).

Pathogenic and indicator organisms die off naturally in wastewater as a result of adverse environmental factors such as temperature, pH, and chemical constituents. Predation, lysis, and parasitism are other possible mechanisms for the natural dying off of these organisms. These factors, coupled with physical removal mechanisms such as sedimentation and filtration, decrease the levels of pathogenic and indicator organisms throughout the various wastewater treatment steps preceding disinfection. Concentration ranges of organisms in domestic wastewater and reduction through primary and secondary treatment are summarized in Tables 3.7, 3.8, and 3.9. Estimated secondary treated effluent concentrations are presented in Chapter 2 (Table 2.3).

Table 3.7   Bacterial densities in domestic wastewater

| Wastewater | Million organisms/100 mL | | | |
	Total coliform	Fecal coliform (FC)	Fecal streptococci (FS)	Ratio FC:FS
"A"	17.2	17.20	4.00	4.3
"B"	33.0	10.90	2.47	4.4
"C"	1.94	0.34	0.064	5.3
"D"	6.30	1.72	0.20	8.6

Recent data (Moore *et al.*, 1994) suggest that the use of coliform bacteria as indicators of protozoan contamination of water may not be adequate. While coliforms were detected in 88% of the waterborne disease outbreaks associated with bacteria, viruses, and unknown agents, coliforms were present in only 33% of the protozoan-linked cases. However, detection of *Giardia* and *Cryptosporidium* is difficult and time consuming and requires concentration of cysts and immunological tests using monoclonal antibodies (Griamason *et al.*, 1994; Musial *et al.*, 1987; and Rose *et al.*, 1989). The lack of sensitivity of the coliform indicator system to pathogens such as parasites suggests the need for a better replacement indicator system to determine disinfection efficacy.

If one assumes that pathogenic organisms are removed in proportion to the indicator organisms (total and/or fecal coliforms), conventional treatment of

Table 3.8   Typical levels of coliform bacteria in domestic wastewater after various wastewater treatment steps (Hubley *et al.*, 1985)

Wastewater	Total coliforms, no./100 mL	Fecal coliforms, no./100 mL
Raw	10^7–10^8	10^6–10^7
Primary effluent	10^7–10^8	10^6–10^7
Secondary	10^5–10^6	10^4–10^5
Filtered secondary	10^4–10^5	10^3–10^5
Nitrified	10^4–10^5	10^3–10^5
Filtered nitrified	10^4–10^5	10^3–10^5

Table 3.9   Microbial reductions by conventional treatment processes (U.S. EPA, 1986)

Microorganism	Primary treatment removal, %	Secondary treatment removal, %
Total coliforms	< 10	90–99
Fecal coliforms	35	90–99
Shigella spp.	15	91–99
Salmonella spp.	15	96–99
E. coli	15	90–99
Viruses	< 10	76–99
Entamoeba histolytica	10–50	10

domestic wastewater without disinfection cannot be considered sufficient for removal and control of human pathogens when the water is to be beneficially reused or body contact will occur (U.S. EPA, 1986).

SURVIVAL OF PATHOGENS IN THE ENVIRONMENT. The potential for exposure to pathogenic organisms diminishes over time because environmental conditions such as heat, sunlight, and desiccation, as well as other microorganisms, destroy pathogens in wastewater and sludge (U.S. EPA, 1992). Table 3.10 summarizes survival rates of four types of pathogenic organisms in soil and on plants (U.S. EPA, 1992). While the disinfection effects of temperature on protozoan cysts have been documented (Bingham *et al.,* 1979, and Fayer, 1994), other studies suggest that the encysted parasites are robust in the environment (Robertson *et al.,* 1992). The threat to public health and animals from protozoa in wastewater and sludge may be high; however, limited long-term data are available. Bacteria, viruses, and

Table 3.10   Survival times of pathogens in soil and on plant surfaces (U.S. EPA, 1992)

Pathogen	Soil		Plants	
	Absolute maximum[a]	Common maximum	Absolute maximum	Common maximum
Bacteria	1 year	2 months	6 months	1 month
Viruses	1 year	3 months	2 months	1 month
Protozoan cysts[b]	10 days	2 days	5 days	2 days
Helminth ova	7 years	2 years	5 months	1 month

[a] Greater survival time is possible under unusual conditions such as consistently low temperatures or highly sheltered conditions (for example, helminth ova below the soil in fallow fields).

[b] Few, if any, data are available on the survival times of *Giardia* cysts and *Cryptosporidium* oocysts.

helminths (particularly helminth eggs [ova], which are the most resistant form in the helminth life cycle) are of much greater concern (U.S. EPA, 1992).

WASTEWATER DISINFECTION TECHNOLOGIES

TYPES OF DISINFECTION TECHNOLOGIES. Pathogens either die naturally or are destroyed in significant numbers in the course of wastewater treatment processes. Such reduction should be distinguished from purposeful disinfection, whereby physical and chemical facilities are installed at WWTPs for inactivating residual pathogens.

Sterilization should not be confused with disinfection. *Sterilization* implies the destruction of all living things in a medium. The production of sterile water is generally confined to research, medical practices, and manufacture of pharmaceuticals (Fair *et al.,* 1968). *Disinfection* is an operation by which living, potentially infectious organisms are killed.

Disinfection can be achieved through the application of heat, light, oxidizing chemicals, acids and alkalies, metal ions, and surface active chemicals (Fair *et al.,* 1968).

Raising water to the boiling point will disinfect it. Because no important waterborne diseases are caused by spore-forming bacteria or other heat-resistant organisms, boiling is a safe and effective practice when drinking water safety is suspect (Fair *et al.,* 1968). However, in wastewater treatment, boiling and other heat-application methods of disinfection are cost prohibitive and not in practice. Heat-generating processes such as thermophilic aerobic digestion, sludge composting, and normal aerobic and anaerobic digestion do decrease pathogens, but these processes are used mainly for purposes other than disinfection. The latest federal rules on sludge disposal and reuse, however, are mandating the use of one or more of these technologies to achieve significant pathogen reduction.

Sunlight is a natural disinfectant, principally acting as a desiccant. Irradiation by UV light intensifies disinfection and makes it a manageable undertaking. The most common source of UV light is a mercury vapor lamp constructed of quartz or a similar material that is transparent to intense and destructive invisible light of 253.7 nm (2 537 Å). The light is emitted by a mercury vapor arc. To ensure disinfection, the water must be free from suspended material and light-absorbing substances. Other forms of radiant or sonic energy are destructive to microorganisms, but they have yet to find engineering application in water disinfection (Fair *et al.,* 1968).

Oxidizing chemicals including the halogens (chlorine, bromine, and iodine), ozone, potassium permanganate, hydrogen peroxide, hypochlorites of sodium and calcium, chlorine dioxide, and bromine chloride can facilitate disinfection if organisms in water or wastewater are exposed to the proper dosage for the appropriate contact time.

Pathogenic bacteria cannot survive long in highly alkaline or highly acidic waters (that is, at high [>11] or low [<3] pH values). Destruction of bacteria during lime softening is an example of pH effects (Fair *et al.,* 1968).

Among surfactants, the cationic detergents are toxic to pathogens, but the anionic detergents only weakly so. Neutral detergents have an intermediate disinfecting strength. As disinfectants, detergents have been applied only selectively in wash and rinse waters of eating establishments (Fair *et al.,* 1968).

While many disinfectants are available, not all have been used for treated wastewater because of cost and other considerations. The appropriate selection of disinfectant is also made difficult by the fact that the susceptibility of different microorganisms to different disinfectants has been found to vary. Chlorination/dechlorination has been the most widely used disinfection technology in the U.S.; ozonation and UV light are emerging technologies. The focus of this manual is on these three methods.

Selection of a disinfection technology for wastewater treatment requires consideration of the various factors outlined in Table 3.11. Table 3.12 lists the applicability of alternative disinfection techniques based on different considerations.

Table 3.11   Major factors in evaluating disinfectant alternatives (U.S. EPA, 1986)

Effectiveness	Ability to achieve target levels of selected indicator organisms
	Broad-spectrum disinfecting ability
	Reliability
Use cost	Capital cost
	Amortization cost
	Operation and maintenance costs
	Cost of special wastewater pretreatment
Practicality	Ease of transport and storage or ease of on-site generation
	Ease of application and control
	Flexibility
	Complexity
	Ability to predict results
	Safety considerations
Pilot studies required	Dose requirements
	Refine design details
Potential adverse effects	Toxicity to aquatic life
	Formation and transmission of undesirable bioaccumulating substances
	Formation and transmission of toxic, mutagenic, or carcinogenic substances

Table 3.12 Applicability of alternative disinfection techniques (U.S. EPA, 1986)

Considerations	Chlorination	Chlorination/ dechlorination	Bromine chloride	Chlorine dioxide	Ozonation	Ultraviolet radiation
Size of plant	All sizes	All sizes	All sizes	Small to medium	Medium to large	Small to medium
Applicable level of treatment prior to disinfection	All levels	All levels	Secondary	Secondary	Secondary	Secondary
Equipment reliability	Good	Fair to good	?[a]	?	Fair to good	Fair to good
Process control	Well developed	Fairly well developed	Problematic	No experience	Fairly well developed	Fairly well developed
Relative complexity of technology	Simple to moderate	Moderate	Moderate	Moderate	Complex	Simple to moderate
Safety concerns	Yes	Yes	Yes	Yes	Yes	No
Transportation on site	Substantial	Substantial	Substantial	Substantial	Minimal	Minimal
Bactericidal	Good	Good	Good	Good	Good	Good
Virucidal	Poor	Poor	Fair to good	Good	Good	Good
Cysticidal	Poor	Poor	?	Fair	Good	Ineffective
Fish toxicity	Toxic	Nontoxic	Slight to moderate	Toxic	None expected	Nontoxic
Hazardous byproducts	Yes	Yes	Yes	Yes	Yes	No
Persistent residual	Long	None	Short	Moderate	None	None

Table 3.12   Applicability of alternative disinfection techniques (U.S. EPA, 1986) (continued)

Considerations	Chlorination	Chlorination/ dechlorination	Bromine chloride	Chlorine dioxide	Ozonation	Ultraviolet radiation
Contact time	Long	Long	Moderate	Moderate to long	Moderate	Short
Contributes dissolved oxygen	No	No	No	No	Yes	No
Reacts with ammonia	Yes	Yes	Yes	No	Yes (high pH only)	No
Color removal	Moderate	Moderate	?	Yes	Yes	No
Increased dissolved solids	Yes	Yes	Yes	Yes	No	No
pH dependent	Yes	Yes	Yes	No	Slight (high pH)	No
Operation and maintenance sensitive	Minimal	Moderate	Moderate	?	High	Moderate
Corrosive	Yes	Yes	Yes	Yes	Yes	No

a ? = unknown.

DISINFECTION TECHNOLOGIES AND MECHANISMS OF MICROBIAL INACTIVATION. Chlorine. Early workers (Green and Stumpf, 1946, and Knox *et al.*, 1948) hypothesized that chlorine inhibited specific enzymes, which ultimately inactivated bacteria (WPCF, 1984). It is now clear that chlorine induces a series of events associated with cell envelope activity (Camper and McFeters, 1979; Haas and Englebrecht, 1980; and Venkobacher *et al.*, 1975 and 1977) and damages nucleic acids, even causing mutations (WPCF, 1984). Inactivation of viruses can be accomplished by damage to either nucleic acids, the viral coat protein, or both (Dennis *et al.*, 1979; Kim and Min, 1979; O'Brien and Newman, 1979; Olivieri *et al.*, 1980; and Tenno *et al.*, 1980).

The effects of different forms of chlorine and the differences in rates of microbial inactivation have been widely researched. Aqueous chlorine at high pH (>8.5) is mostly in the form of OCl⁻, which is less potent than the non-ionized form (HOCl [hypochlorous acid]) predominant in low pH (less than pH 6.5) wastewater. Both HOCl and OCl⁻ are known as free chlorine. Chlorine reacts with ammonia in water and wastewater to form chloramines (mono-, di-, and trichloramines) or combined chlorines, which are also disinfectants of various potencies but are less potent than HOCl. Organic nitrogen (including amino acids and proteins) is also known to react with chlorine, forming organochloramines, the potency of which is debated. However, these are generally considered to have no practical disinfection value. Researchers have also found that different forms of chlorine may be effective for different pathogens.

While chlorine added to water at adequate and consistent levels can be relatively effective for disinfection of *Giardia* (Hoff, 1986), it is not as effective for *Cryptosporidium* (Sterling, 1990). Korich *et al.* (1990) found that monochloramine and chlorine, even at 80 ppm, required 90 minutes of contact time to achieve 90% inactivation of *Cryptosporidium*. Chlorine dioxide (1.3 ppm) achieved 90% inactivation after 1 hour. The addition of chlorine compounds would not be acceptable as the sole method to remove protozoan cysts from drinking water. Filtration is required (Moore *et al.*, 1994).

Ozone. The literature suggests that ozone inactivates bacteria by totally or partially destroying the cell wall; this is followed by lysis of the cell (WPCF, 1984). Scott (1974) showed that the attack of ozone on purines and pyrimidines found in the nucleic acids results in addition of OH and hydrogen across the double bonds, opening of the rings, or formation of thymine dimers. Ozone has also been shown to break nitrogen–carbon bonds between sugars and the bases and to affect DNA by breaking hydrogen bonds. Depolymerization and the breaking of phosphate sugar bonds have resulted from these reactions. Ozone has also been shown to induce chromosome breakage and deletions. Scott concluded that the primary reaction of ozone was with the unsaturated fatty acids of the phospholipids, glycoproteins, and glycolipids in the cell membrane, resulting in leakage of cellular constituents outside the cell. Perrich *et al.* (1975) reported similar results on the leakage of cell contents and hypothesized irreversible enzyme inhibition (WPCF, 1984).

The mechanisms of virus inactivation are as poorly understood as those of bacterial inactivation. Work by Riesser *et al.* (1976) indicated that disinfection with ozone destroys the viral proteins so that invasion of a susceptible cell is prohibited. Perrich *et al.* (1975) have also suggested the alteration of the viral capsid as one of two possible mechanisms; the other is the inactivation of viral RNA (WPCF, 1984).

Other studies have attempted to determine whether viruses are inactivated by ozone coagulation of the protein capsid or by oxidative disruption of the nucleic acid core (Kim *et al.*, 1980, and Shinriki *et al.*, 1980). Shinriki *et al.* (1980) studied the RNA of the tobacco mosaic virus after exposure to ozone and concluded that the most important factor in virus inactivation may be ozone's preferential attack on the guanine base (WPCF, 1984).

Ozone is a more effective disinfectant for *Cryptosporidium* than chlorine compounds. Ozone (1 ppm) achieved greater than 90% inactivation of oocysts after a 5-minute contact time (Korich *et al.*, 1990). Unfortunately, no residual is provided with ozone. Korich *et al.* (1990) calculated that *C. parvum* oocysts are 30 times more resistant to ozone disinfection than *Giardia*. Finch *et al.* (1993) achieved a 3-log inactivation of *Giardia* with approximately 1.25 mg ozone min/L at 22°C. However, addition of ozone is not considered acceptable as the sole means of removing protozoan cysts from drinking water. Filtration is required (Moore *et al.*, 1994).

Ultraviolet Radiation. The primary mechanism of UV light in inactivating microorganisms is direct damage of the cellular nucleic acids (Bridges, 1976, and Setlow, 1965). When UV energy is absorbed by the genetic material of microorganisms, pyrimidine dimers are formed and join neighboring cytosine or thymine moieties by a cyclobutane ring (Bridges, 1976). These dimers are the major cause of the lethal and mutagenic effects of UV radiation. The pyrimidine dimers prevent DNA from replicating, which results in death of the cell (Witkin, 1976).

Ultraviolet light at a wavelength of 265 nm causes the most cellular damage (Gates, 1929, and Nagy, 1964). The wavelength coincides closely with the absorption maximum of nucleic acids. Low-pressure mercury discharge lamps emit approximately 92% of their light at a wavelength of 254 nm (Nagy, 1964, and Yip and Konasewich, 1972). Thus, these lamps are nearly ideal UV light generators (WPCF, 1984).

When injured organisms are exposed to light energy at wavelengths between 310 and 500 nm, fission of pyrimidine dimers occurs, the original base sequence is restored, and the organisms can replicate normally. This phenomenon is called photoreactivation and was discovered in 1949, independently, by Kelner (1949) and Dulbecco (1949). Reactivation is never complete, and only a fraction of the affected organisms recover (Lamanna *et al.,* 1973). The secondary light treatment reduces the UV inactivation by a constant factor. The degree of reactivation, which is mediated by a single enzyme, is proportional to the time, intensity of exposure, and temperature. Recovery from UV inactivation can also occur in the dark. Dark reactivation requires much longer times and involves two distinct enzyme systems (WPCF, 1984).

MICROBIAL INACTIVATION UNRELATED TO DISINFECTION.

Aside from disinfection, microbial inactivation occurs through dilution, physical removal, and natural die-off. Mechanisms for natural die-off may include such factors as sunlight, temperature, salinity, heavy metals, bacteriophage, parasitism, lysis, and predation (Jones, 1971). The impacts of dilution on the concentration of organisms in wastewater discharged to a stream, river, or lake can be estimated. Several mathematical models have been proposed for estimating natural bacterial die-off, and these models are primarily based on first-order kinetics (U.S. EPA, 1986). For example, the first-order model for die-off in streams is as follows (Hubley *et al.,* 1985):

$$N = N_o e^{(-kt)} \tag{3.1}$$

Where

N_o =   the initial concentration of microorganisms discharged into the stream,

N =   the concentration of the microorganisms at time t after discharge to the stream,

k =   rate constant, and

t =   elapsed time.

The first-order model for die-off in standing water bodies (for example, lagoons and lakes) is as follows (U.S. EPA, 1986):

$$N = N_o / (1 + kt_d) \tag{3.2}$$

Where

t_d =   the hydraulic detention time in the water body based on the water body's discharge.

The rate constant k can be determined from a die-away study of typical lakes or streams in a planning area. Rate constants may also be found in the literature (Berg, 1978).

Johnson *et al.* (1979) found that coliform die-away or removal rates can be as much as 16 times higher in summer months than in winter months. This could be attributed to low metabolic activity of the organisms at lower

---

temperatures. Results of a study on the Logan, Utah, lagoon system indicated that the summer coliform decay rate coefficient was 0.5/day and the winter coliform decay rate coefficient was 0.03/day (U.S. EPA, 1986).

DISINFECTION KINETICS. Dose, Residual, Demand, and Contact Time. When natural die-off is not sufficient to prevent the potential for humans to ingest pathogenic organisms, disinfection should be required before discharge. Disinfection is a time-dependent process. The outcome of bacterial and viral destruction is the result of a series of physical, chemical, and biochemical actions that can be approximated by simple kinetic expressions. Although the kinetic descriptions are simple, applications are not universal. Site-specific conditions may create problems with precision and accuracy in the use of an empirical relationship that is effective at another site (U.S. EPA, 1986).

In addition to time (often referred to as *contact time*), disinfection also depends on the intensity or amount of the physical (as in UV light) or chemical (as in the case of the oxidants) entity applied to water and wastewater for the inactivation of the microorganisms. This amount is known as the "dose." In the case of chemical disinfection, because the chemicals used are often oxidants, a portion of the dosed chemicals is usually consumed in reactions with reducing agents present in wastewater. Such reactions are typically fast. Hence, the full amount of the applied dose is not available for disinfection. The part of the chemical dose that is consumed in the extraneous chemical reactions is known as the "demand." The part of the dosed chemical left after satisfaction of the demand and available for disinfection is known as the "residual." Thus, the kinetic expressions for microbial inactivation should be based on the residual and not the applied dose. In UV disinfection, the concepts of residual and demand do not apply. In chlorine disinfection of wastewater, the concepts of dose, residual, and demand are complicated by the fact that chlorine reacts with ammonia, resulting in other disinfectants called *combined chlorine* (for example, monochloramine). Such reactions are known to depend on the ratio of the chlorine dose to ammonia. The dose/residual relationship is known to be nonlinear and is described by the famous "breakpoint curve." The kinetics of chlorine disinfection are also complicated by the fact that various species of chlorine resulting from the reactions with ammonia have varying inactivation potentials for the microorganisms.

Kinetic Models of Disinfection. The information needed for designing a disinfection system includes knowledge of the rate of inactivation of the target organism(s) by the disinfectant. In particular, the effect of disinfectant concentration on the disinfection rate will determine the most efficient combination of contact time and disinfectant dose to use. The major precepts of disinfection kinetics were first enunciated by Chick (1908), who recognized the similarity between microbial inactivation by chemical disinfectants and chemical reactions (U.S. EPA, 1986). Chick postulated that the rate of disinfection could be described in the following manner:

$$-\frac{dN}{dt} = kN \qquad (3.3)$$

Where

$-\dfrac{dN}{dt}$ = rate of decrease in organism population,

k = organism die-off rate constant, and

N = number of surviving organisms at any given time t.

The solution of this equation follows first-order kinetics similarly to Equation 3.1 and is often referred to as *Chick's law*. Divergences from exponential decay in disinfection are commonly observed, and it is recognized that many factors can cause these deviations (for example, changes in disinfectant concentration with time, differences in resistance between individual organisms of various ages in the same culture, existence of clumps of organisms, or occlusion of organisms by suspended solids) (U.S. EPA, 1986).

Watson (1908) analyzed data with varying concentrations of disinfectant and demonstrated a definite logarithmic relationship between the concentration of disinfectant and the mean reaction velocity. He proposed the following equation to relate the rate constant of inactivation to the disinfectant concentration (U.S. EPA, 1986):

$$k = k'C^n \qquad (3.4)$$

Where

C = disinfectant concentration;

n = coefficient of dilution; and

k' = corrected die-off rate constant, presumably independent of C and N, the number of organisms.

The combination of Equations 3.3 and 3.4 yields

$$-\frac{dN}{dt} = k'NC^n \qquad (3.5)$$

The process of disinfection is influenced by temperature, and the Arrhenius equation can be used to predict temperature effects when direct heat-kill is not a significant factor (U.S. EPA, 1986):

$$k'_T = k'_{20}B^{(T-20)} \qquad (3.6)$$

Where

k'_T = rate constant at temperature T, °C;

k'_{20} = rate constant at 20°C; and

B = empirical constant related to activation energy and universal gas constant.

Little is known about disinfection efficiency at elevated temperatures, but for agents such as ozone, a significant reduction would occur because of the lower efficiency of ozone mass transfer and greater ozone decay (U.S. EPA, 1986).

The observation has often been made that inactivation of organisms in batch experiments, even when the disinfectant concentration is kept constant, does not follow the exponential decay pattern predicted by Equation 3.3 (U.S. EPA, 1986). Various attempts have been made to refine Chick's law or Chick-Watson models. Hom (1972) developed a flexible but highly empirical kinetic formulation based on the modification of Equations 3.3 and 3.4 to the following form:

$$-\frac{dN}{dt} = k'Nt^m C^n \qquad (3.7)$$

Where m is an empirical constant.

For changing concentrations of disinfectant, the observed disinfection efficiency is generally approximated by the following relationship (Fair et al., 1968):

$$C^n t_p = \text{constant} \qquad (3.8)$$

Where

n = coefficient of dilution or a measure of the order of the reaction, and

t_p = time required to produce a constant percentage of die-off.

This observation has evolved into a "CT" concept that is currently being used in water treatment regulations to ensure a certain percentage die-off of Giardia, viruses, Cryptosporidium, and other organisms. The percentage kills are being expressed in terms of log removals.

Collins et al. (1971) developed a model based on a comprehensive pilot plant study of a primary effluent. Their original model was subsequently refined (Collins et al., 1974), based on plant-scale studies, as the following (White, 1992):

$$N = N_o(1 + 0.23Ct)^{-3} \qquad (3.9)$$

Further refinement of the Collins model was presented later as follows (White, 1992):

$$N = N_o(Ct/b)^{-n} \qquad (3.10)$$

Where b and n are empirical constants.

A model developed by Selleck et $al.$ (1978) is shown in Equation 3.11 (WPCF, 1984):

$$\log\ (N/N_o) = -a[\log(1 + Ct/b)] \qquad (3.11)$$

Where a is a constant. Note that while N, N_o, C, and t are common terms, the similar empirical constants used in different equations above may not be related to each other. In all of the equations above, the constants are functions of the nature of the wastewater, chemical species, and, in the case of chlorine, the chlorine dose-to-ammonia ratio, all of which limit the usefulness of the models (WPCF, 1984). While most of the models presented were developed for chlorination, these may also be applicable to other oxidants. Further discussions of the kinetic models specific to the various disinfectants are provided in Chapters 4, 5, and 6.

Regrowth Phenomenon. The kinetics of disinfection are also complicated by the fact that disinfectant-injured organisms, through repair mechanisms, can reactivate to a limited extent both in light and dark. This reactivation, sometimes referred to as *regrowth,* is mostly reported in the case of UV disinfection. As the reactivation is never complete and only a fraction of the affected organisms recover (Lamanna *et al.,* 1973), the problem of reactivation may not be severe.

The protozoan parasites *Cryptosporidium* and *Giardia* are found outside their animal hosts as oocysts and cysts, respectively. Once inside the appropriate host, they excyst and form the infectious trophozoites that infect the gastrointestinal tract and reproduce. Currently, it is not clear how enumeration of the oocysts and cysts relates to their viability or potential infectivity. However, determination of cyst viability is difficult, requiring *in vitro* excystation (Haas *et al.,* 1994).

Plant Control of Disinfection. Control of disinfection at a WWTP must account for the different variables affecting disinfection. Though a universal kinetic model of disinfection is not yet available, it is clear that the main control variables for disinfection are disinfectant residual (or dose as applicable) and contact time needed to achieve a certain effluent concentration of microorganisms. The contact time is normally fixed through design and construction of the disinfection facility and the flow rates to the facility. The residual (or dose in the case of UV disinfection) is the main control variable for a disinfection facility. In the case of chlorination, both manual and automatic controls of residual are available. For automatic control, different control algorithms, such as flow-paced, residual trim, feed-forward, feed-back, and compound loop controls, are used.

Some researchers maintain that in the case of disinfection with oxidants, oxidation-reduction potential (ORP), and not the oxidant residual, should be used as the control variable for disinfection. The proponents of this control approach maintain that because different disinfectants have different disinfecting powers, relating the microorganism inactivation with the residual concentration can be unreliable. They maintain that the bacterial kill rate depends on ORP irrespective of the residual levels. Increasing support for their contention has been developing in the industry because of the fact that different oxidants have different disinfecting powers, probably because of their differing oxidation potentials. For example, ozone (though expensive) has a higher oxidation potential than any of the chlorine residual species. In chlorination, an un-ionized fraction of free chlorine has a higher oxidation potential (and a higher disinfecting power) than its ionized counterpart. Combined inorganic chlorines have lower oxidation potentials (and lower disinfection strengths) than free chlorine. Where a mixture of these various chlorine residual species can occur, total residual of differing speciation can ideally result in different levels of microbial inactivation. Oxidation-reduction potential may thus be able to serve as a combined surrogate parameter for disinfection strength. Use of the ORP method of control for disinfection has so far been limited by the absence of a reliable instrument for its measurement and by the fact that different wastewaters may have different background ORPs based on the presence of other extraneous oxidants (like dissolved oxygen) and reductants (like organic matter) in wastewater. Some manufacturers of ORP instruments claim that instrument reliability in ORP measurements has been achieved. The use of such instruments for both chlorination and dechlorination has been successfully tried in many WWTPs around the country.

*R*ESIDUAL TOXICITY OF DISINFECTANTS

Wastewater treatment plants should maintain a certain residual in the wastewater to ensure bacterial die-off during chemical disinfection. Because disinfection is typically the last process in a WWTP before discharge, the residual gets transferred with the treated discharge. Residuals of the oxidizing disinfectants have long been believed to be toxic to aquatic life including fish in the receiving body of water. Information from researchers in the water treatment industry also indicate that some of the disinfection processes produce hazardous byproducts that are suspected of being carcinogens (such as trihalomethanes) or cause other complications to humans. Ultraviolet disinfection, which does not leave a residual, is commonly believed to be nontoxic to aquatic biota. Lately, regulatory agencies have been mandating limits on acute and chronic toxicities of effluent in some test organisms. Beside these parameters, limits are also imposed on the residual levels of oxidants (such as chlorine) and some hazardous byproducts (such as total trihalomethanes).

A comparative analysis of various disinfection technologies with respect to fish toxicity, generation of hazardous byproducts, the persistence of residuals, and other considerations are presented in Table 3.12. However, most of the available research and data on toxicity apply to chlorine; data for other disinfectants are sparse.

A review of the literature on the toxicity and environmental impact of disinfection was presented in *Wastewater Disinfection, A State-of-the-Art Report* (WPCF, 1984), some of which is summarized below. Further details on the residual toxicity of disinfectants are available in Chapters 4, 5, and 6 of this manual.

Trout, salmon, and some other organisms tend to be more sensitive to chlorine than other freshwater animals (U.S. EPA, 1976). A literature review of data on warm and cold freshwater and marine fish suggested no-adverse-effect levels of 0.01, 0.002, and 0.01 mg/L total residual chlorine (WPCF, 1984). Dechlorination reduced or eliminated toxicity; brominated effluents were at least as toxic as, but had more labile residuals than, chlorinated effluents.

A Canadian group noted the possibility of behavioral alterations among fish populations in the presence of a chlorine residual (Larrick *et al.*, 1978). They tested golden shiner avoidance to chlorinated river water and hypothesized an avoidance threshold of 0.015 to 0.017 mg/L of HOCl as chlorine, regardless of temperature. Bogardus *et al.* (1978) found alewives to have an avoidance threshold of less than 0.03 mg/L of monochloramine as chlorine, while coho salmon had a threshold of greater than 0.2 mg/L (WPCF, 1984).

Block *et al.* (1978) exposed adult white perch to 0.8 mg/L total residual oxidant, dosed either as chlorine or ozone in an estuary. No fish died in the chlorinated system until after 8 hours, while all fish in the ozonated system died within 6 hours. The type of chemical species is an important factor determining the toxic responses in an ozonated estuarine system (WPCF, 1984).

Asbury and Coler (1980) monitored the toxic effects of ozone residual in river water on eggs and larvae of yellow perch, fathead minnows, common suckers, and bluegill sunfish. Fathead minnow larvae seemed to be the most sensitive, having an LC_{50} (concentration resulting in death of 50% of population) of less than 0.1 mg/L at exposure times as short as 30 seconds (WPCF, 1984).

Virtually no data exist with which to evaluate the environmental impact of chlorine dioxide or UV radiation (WPCF, 1984).

The following were some of the conclusions reached in a literature search (WPCF, 1984):

- Chlorine residual levels may need to be as low as 0.002 mg/L to preclude adverse effects.
- The minimal data available on toxic effects of ozone residuals indicate that ozone and chlorine are of similar toxicity; however, the greater instability of ozone would tend to reduce impacts.
- Dechlorination can virtually eliminate toxic effects resulting from wastewater chlorination.

*R*EFERENCES

Akin, E.W., *et al.* (1978) Health Hazards Associated with Wastewater Effluents and Sludge: Microbiological Considerations. In *Proc. Conf. Risk Assessment Health Eff. Land Appl. Munic. Wastewater and Sludges.* B.P. Sagik and C.A. Sorber (Eds.), Univ. of Texas, San Antonio.

American Public Health Association (1995) *Standard Methods for the Examination of Water and Wastewater.* 19th Ed., Washington, D.C.

Asbury, C., and Coler, R. (1980) Toxicity of Dissolved Ozone to Fish Eggs and Larvae. *J. Water Pollut. Control Fed.,* **52**, 1990.

Banatvala, J.E. (1981) Viruses in Feces. In *Proc. Conf. Viruses Wastewater Treatment.* M. Goddard and M. Butler (Eds.), Pergamon Press, Oxford, U.K.

Berg, G. (1978) *Indicators of Viruses in Water and Food.* Ann Arbor Science, Ann Arbor, Mich.

Bingham, A.K., *et al.* (1979) Induction of *Giardia* Excystation and the Effect of Temperature on Cyst Viability as Compared by Eosin-exclusion and In Vitro Excystation. In *Waterborne Transmission of Giardiasis.* W. Jakubowski and J.C. Hoff (Eds.), EPA-600/9-79-001, U.S. EPA, Cincinnati, Ohio.

Block, R.M., *et al.* (1978) Respiratory and Osmoregulatory Responses of White Perch (Monrone Americana) Exposed to Chlorine and Ozone in Estuarine Waters. In *Water Chlorination: Environmental Impact and Health Effects.* Vol. 2, R.L. Jolley *et al.* (Eds.), Ann Arbor Science, Ann Arbor, Mich.

Bogardus, R.B., *et al.* (1978) Avoidance of Monochloramine: Test Tank Results for Rainbow Trout, Coho Salmon, Alewife, Yellow Perch, and Spottail Shiner. In *Water Chlorination: Environmental Impact and Health Effects.* Vol. 2, R.L. Jolley *et al.* (Eds.), Ann Arbor Science, Ann Arbor, Mich.

Bridges, B.A. (1976) Survival of Bacteria Following Exposure to Ultraviolet and Ionizing Radiations. In *The Survival of Vegetative Microbes.* T.R.G. Gray and J.R. Posgate (Eds.), Cambridge University Press, U.K.

Bryan, F.L. (1974) Diseases Transmitted by Foods Contaminated by Wastewater. In *Proc. Wastewater Use Prod. Food and Fiber.* Public Health Serv., U.S. Dep. Health, Educ., Welfare, Washington, D.C.

Cabelli, V.J. (1977) Indicators of Recreational Water Quality. In *Bacterial Indicators/Health Hazards Associated with Water.* A.W. Hoadley and B.J. Dutka (Eds.), Technical Publication 635, ASTM, Philadelphia, Pa.

Cabelli, V.J. (1980) Health Effects Quality Criteria for Marine Recreational Waters. EPA-600/1-80-031, U.S. EPA, Cincinnati, Ohio.

Cabelli, V.J., *et al.* (1975a) Relationship of Microbial Indicators to Health Effects at Marine Bathing Beaches. Paper presented at Annu. Meeting Am. Public Health Assoc., Chicago, Ill.

Cabelli, V.J., *et al.* (1975b) The Development of Criteria for Recreation Water. In *Discharge of Sewage from Sea Outfalls.* A.L.H. Gameson (Ed.), Pergamon Press, New York, N.Y.

Cabelli, V.J., *et al.* (1976) The Impact of Pollution on Marine Bathing Beaches: An Epidemiological Study. *Soc. Limnol. Oceanogr. Spec. Symp.,* **2**, 424.

Camper, A.K., and McFeters, G.A. (1979) Chlorine Injury and the Enumeration of Waterborne Coliform Bacteria. *Appl. Environ. Microbiol.,* **37**, 633.

Casson, L.W., *et al.* (1992) HIV Survivability in Wastewater. *Water Environ. Res.,* **64**, 213.

Chick, H. (1908) An Investigation of the Laws of Disinfection. *J. Hyg.,* **8**, 92.

Collins, H.F., *et al.* (1971) Problems in Obtaining Adequate Sewage Disinfection, *J. Sanit. Eng. Div., Proc. Am. Soc. Civ. Eng.,* **97**, SA5.

Collins, H.F., *et al.* (1974) *Interim Manual for Wastewater Chlorination and Dechlorination Practices.* Calif. State Dep. Health, Sacramento.

Comptroller General of the United States (1977) *Unnecessary and Harmful Levels of Domestic Sewage Chlorination Should be Stopped.* CED-77-108, Rep. to Congress, U.S. Gen. Accounting Office, Washington, D.C.

Curds, C.R. (1992) *Protozoa in the Water Industry.* Cambridge University Press, Cambridge, U.K.

Dennis, W.H., *et al.* (1979) Mechanism of Disinfection: Incorporation of ^{36}Cl into f2 Virus. *Water Res.,* **13**, 363.

Dulbecco, R. (1949) Reactivation of Ultraviolet-Inactivated Bacteriophage by Visible Light. *Nature,* **163**, 949.

DuPont, H.L., *et al.* (1995) The Infectivity of *Cryptosporidium parvum* in Healthy Volunteers. *N. Engl. J. Med.,* **332**, 13.

Fair, G.M., *et al.* (1968) *Water and Wastewater Engineering. Vol. 2: Water Purification and Wastewater Treatment and Disposal.* John Wiley & Sons Inc., New York, N.Y.

Farzadegan, H. (1991) *Proceedings of a Symposium: Survival of HIV in Environmental Waters.* Natl. Sci. Foundation, Johns Hopkins Univ. School Hyg. Public Health, Baltimore, Md.

Fayer, R. (1994) Effect of High Temperature on Infectivity of *Cryptosporidium parvum* Oocysts in Water. *Appl. Environ. Microbiol.,* **60**, 8.

Fayer, R., and Ungar, B.L.P. (1986) *Cryptosporidium* spp. and Cryptosporidiosis. *Microbiol. Rev.,* **50**, 458.

Finch, F.R., *et al.* (1993) Comparison of *Giardia lamblia* and *Giardia muris* Cyst Inactivation by Ozone. *Appl. Environ. Microbiol.,* **59**, 3674.

Gates, F.L. (1929) A Study of the Bactericidal Action of Ultraviolet Light II. The Effects of Various Environmental Factors and Conditions. *J. Gen. Physiol.,* **13**, 249.

Gaudy, A.F., and Gaudy, E.T. (1980) *Microbiology for Environmental Scientists and Engineers.* McGraw-Hill, Inc., New York, N.Y.

Geldreich, E.E. (1966) Sanitary Significance of Fecal Coliforms in the Environment. WP-20-3, FWPCA, Robert A. Taft Sanit. Eng. Center, U.S. Dep. Int., Cincinnati, Ohio.

Green, D.E., and Stumpf, P.K. (1946) The Mode of Action of Chlorine. *J. Am. Water Works Assoc.,* **38**, 1301.

Griamason, A.M., *et al.* (1994) Application of DAPI and Immunofluorescence for Enhanced Identification of *Cryptosporidium* spp. Oocysts in Water Samples. *Water Res.,* **28**, 733.

Haas, C.N., and Englebrecht, R.S. (1980) Physiological Alterations of Vegetative Microorganisms Resulting from Aqueous Chlorination. *J. Water Pollut. Control Fed.,* **52**, 1976.

Haas, C.N., *et al.* (1994) A Volumetric Method for Assessing *Giardia* Inactivation. *J. Am. Water Works Assoc.,* **86**, 115.

Hoff, J.C. (1986) Inactivation of Microbiological Agents by Chemical Disinfectants. EPA-600/S2-86-067, U.S. EPA, Cincinnati, Ohio.

Hom, L.W. (1972) Kinetics of Chlorine Disinfection in an Eco-System. *J. Sanit. Eng. Div., Proc. Am. Soc. Civ. Eng.*, **98**, 183.

Hubley, D., *et al.* (1985) Risk Assessment of Wastewater Disinfection. EPA-600/2-85-037, U.S. EPA, Cincinnati, Ohio.

Johnson, B.A, *et al.* (1979) Wastewater Stabilization Lagoon Microorganism Removal Efficiency and Effluent Disinfection with Chlorine. EPA-600/2-79-018, U.S. EPA, Cincinnati, Ohio.

Johnson, R.W., *et al.* (1994) HIV and the Bloodborne Pathogen Regulation: Implications for the Wastewater Industry. *Water Environ. Res.,* **66**, 5.

Jones, G.E. (1971) The Fate of Freshwater Bacteria in the Sea. In *Developments in Industrial Microbiology.* Vol. 12, Am. Inst. Biol. Sci., Washington, D.C.

Kelner, A. (1949) Effects of Visible Light on the Recovery of Streptomyces griseus Conidia from Ultraviolet Irradiation Injury. *Proc. Natl. Acad. Sci.,* **35**, 73.

Kim, C.K., and Min, K.H. (1979) Inactivation of Bacteriophage f2 with Chlorine. *Misaengmul Hakhoe Chi,* **16**, 2, 62.

Kim, C.K., *et al.* (1980) Mechanisms of Ozone Inactivation of Bacteriophage f2. *Appl. Environ. Microbiol.,* 39, 210.

Knox, W.W., *et al.* (1948) The Inhibition of Sulfhydryl Enzymes as the Basis of the Bactericidal Action of Chlorine. *J. Bacteriol.,* **55**, 451.

Korich, D.G., *et al.* (1990) Effects of Ozone, Chlorine Dioxide, Chlorine and Monochloramine on *Cryptosporidium parvum* Oocyst Viability. *Appl. Environ. Microbiol.,* **56**, 1423.

Lamanna, C., *et al.* (1973) *Basic Bacteriology.* Williams & Wilkins, Baltimore, Md.

Larrick, S.R., *et al.* (1978) The Use of Various Avoidance Indices to Evaluate the Behavioral Response of the Golden Shiner to Components of Total Residual Chlorine. In *Water Chlorination: Environmental Impact & Health Effects.* Vol. 2, R.L. Jolley *et al.* (Eds.), Ann Arbor Science, Ann Arbor, Mich.

LeChevallier, M.W., *et al.* (1991) Occurrence of *Giardia* and *Cryptosporidium* spp. in Surface Water Supplies. *Appl. Environ. Microbiol.,* **57**, 2610.

McDougal, J.S., *et al.* (1985) Immunoassay for the Detection and Quantification of Infectious Human Retrovirus, Lymphadepathy-Associated Virus (LAV). *J. Immunol. Methodol.,* **76**, 171.

Metcalf & Eddy, Inc. (1991) *Wastewater Engineering, Treatment, Disposal, Reuse.* 3rd Ed., McGraw-Hill, Inc., New York, N.Y.

Moore, A.C., *et al.* (1994) Waterborne Disease in the United States, 1991 and 1992. *J. Am. Water Works Assoc.,* **86**, 87.

Moore, B. (1954a) A Survey of Beach Pollution at a Seaside Resort. *J. Hyg. (G.B.),* **52**.

Moore, B. (1954b) Sewage Contamination of Coastal Bathing Waters. *Bull. Hyg. London,* **29**, 689.

Musial, C.E., *et al.* (1987) Detection of *Cryptosporidium* in Water by Using Polypropylene Cartridge Filters. *Appl. Environ. Microbiol.,* **53**, 687.

Nagy, R. (1964) Application and Measurement of Ultraviolet Radiation. *Am. Ind. Hyg. Assoc. J.,* **25**, 274.

O'Brien, R.T., and Newman, J. (1979) Structural and Compositional Changes Associated with Chlorine Inactivation of Polioviruses. *Appl. Environ. Microbiol.,* **38**, 1034.

Olivieri, V.P., *et al.* (1980) Reaction of Chlorine and Chloramines with Nucleic Acids under Disinfection Conditions. In *Water Chlorination: Environmental Impact and Health Effects.* Vol. 3, R.L. Jolley *et al.* (Eds.), Ann Arbor Science, Ann Arbor, Mich.

Ongerth, J.E., and Stibbs, H.H. (1987) Identification of *Cryptosporidium* Oocysts in River Water. *Appl. Environ. Microbiol.,* **53**, 672.

Perrich, J., *et al.* (1975) Inactivation Kinetics of Viruses and Bacteria in a Model Ozone Contacting Reactor System. *Proc. 2nd Int. Symp. Ozone Technol.,* Int. Ozone Inst.

Riesser, V., *et al.* (1976) Possible Mechanisms of Poliovirus Inactivation by Ozone. Paper presented at the Forum on Ozone Disinfection, Chicago, Ill.

Riggs, J.L. (1989) AIDS Transmission in Drinking Water: No Threat. *J. Am. Water Works Assoc.,* **81**, 69.

Robertson, L.J., *et al.* (1992) Survival of *Cryptosporidium parvum* Oocysts Under Various Environmental Pressures. *Appl. Environ. Microbiol.,* **58**, 11.

Rose, J.B., *et al.* (1989) Evaluation of Immunofluorescence Techniques for Detection of *Cryptosporidium* Oocysts and *Giardia* Cysts from Environmental Samples. *Appl. Environ. Microbiol.,* **55**, 3189.

Rose, J.B., *et al.* (1991) Survey of Potable Water Supplies for *Cryptosporidium* and *Giardia. Environ. Sci. Technol.,* **26**, 1393.

Rosenberg, M.S., *et al.* (1975) Transmission of Shigellosis by Swimming in a Contaminated River. Public Health Serv., Cent. Dis. Control, Atlanta, Ga.

Sabin, A.B. (1957) Properties of Attenuated Poliovirus and Their Behavior in Human Beings. In *Cellular Biology, Nucleic Acids and Viruses.* Vol. 5, New York Academy of Sciences, New York, N.Y.

Scott, D. (1974) The Effect of Ozone on Nucleic Acids and Their Derivatives. In *Aquatic Applications of Ozone.* W.J. Blogoslawski and R.G. Rice (Eds.), Ozone Press International, Syracuse, N.Y.

Selleck, R.E., *et al.* (1978) Kinetics of Bacterial Deactivation with Chlorine. *J. Environ. Eng.,* **104**, 1197.

Setlow, J.K. (1965) The Molecular Basis of Biological Effects of Ultraviolet Radiation and Photoreactivation. *Curr. Topics Radiol. Res.,* **2**, 195.

Shinriki, N., *et al.* (1980) Some Considerations on the Mechanism of Virus Inactivation with Ozone. Paper presented at 4th World Ozone Congress, Int. Ozone Assoc., Houston, Tex.

Sobsey, M., and Olsen, B. (1983) Microbial Agents of Waterborne Disease. In *Assessment of Microbiology and Turbidity Standards for Drinking Water.* EPA-570/4-83-001, U.S. EPA, Washington, D.C.

Spire, B., *et al.* (1984) Inactivation of LAV by Chemical Disinfectants. *Lancet (G.B.),* **2**, 899.

Spire, B., *et al.* (1985) Inactivation of LAV by Heat, Gamma Rays and Ultraviolet Light. *Lancet (G.B.),* **1**, 188.

Sterling, C.R. (1990) Waterborne Cryptosporidiosis. In *Cryptosporidiosis of Man and Animals.* J.P. Dubey, *et al.* (Eds.), CRC Press, Boca Raton, Fla.

Stevenson, A.H. (1953) Studies of Bathing Water Quality and Health. *Am. J. Public Health,* **43**, 529.

Tenno, K.M., *et al.* (1980) The Mechanism of Inactivation of Poliovirus by Hypochlorous Acid. In *Water Chlorination: Environmental Impact and Health Effects.* Vol. 3, R.L. Jolley *et al.* (Eds.), Ann Arbor Science, Ann Arbor, Mich.

U.S. Environmental Protection Agency (1976) *Disinfection of Wastewater.* EPA-430/9-75-012, Washington, D.C.

U.S. Environmental Protection Agency (1986) *Municipal Wastewater Disinfection Design Manual.* EPA-625/1-86-021, Cincinnati, Ohio, 247.

U.S. Environmental Protection Agency (1992) *Control of Pathogens and Vector Attraction in Sewage Sludge.* EPA-625/R-92-013, Environ. Reg. Technol., Office Res. Dev., Washington, D.C.

Venkobacher, C., *et al.* (1975) Mechanism of Disinfection. *Water Res. (G.B.),* **9**, 119.

Venkobacher, C., *et al.* (1977) Mechanism of Disinfection: Effect of Chlorine on Cell Membrane Functions. *Water Res. (G.B.),* **11**, 727.

Water Pollution Control Federation (1984) *Wastewater Disinfection, A State-of-the-Art Report.* Washington, D.C., 78.

Water Pollution Control Federation (1991) *Biological Hazards at Wastewater Treatment Facilities.* Special Publication, Alexandria, Va.

Watson, H.E. (1908) A Note on the Variation of the Rate of Disinfection with Change in the Concentration of the Disinfectant. *J. Hyg. (G.B.),* **8**, 536.

White, G.C. (1992) *Handbook of Chlorination and Alternative Disinfectants.* 3rd Ed., Van Nostrand Reinhold, New York, N.Y.

Witkin, E. (1976) Ultraviolet Mutagenesis and Inducible DNA Repair in *Escherichia coli. Bacteriol. Rev.,* **40**, 869.

Yip, R.W., and Konasewich, D.E. (1972) Ultraviolet Sterilization of Water—Its Potential and Limitations. *Water Pollut. Control,* **110**, 6, 14.

Chapter 4
Reactor Dynamics

62	Introduction	84	Example 2: Pulse-Feed Test
62	Ideal Reactors	84	Data Collection
65	Tracer Test Methods	87	Tracer Results
65	Wastewater Flow Conditions	87	Data Analysis
66	Tracer Selection	87	Data Interpretation
67	Tracer Injection and Detection	88	Combining Tracer Analysis and
69	Tracer Curves		Disinfection Kinetics
69	*E* Curve	90	Disinfection Kinetics
71	*C* Curve	91	Ideal Models
71	*F* Curve	91	Using the Segregated Model
73	Mean Residence Time	92	Using the Tanks-in-Series Model
74	Variance	92	Using the Dispersion Model
75	Nonideal Reactors	92	Calculating a Reactor's Potential
75	Typical Wastewater Disinfection		Efficiency
	Reactors	93	Example 3: Disinfection
76	Conceptual Models of Nonideal		Efficiency Estimation
	Reactors	93	Data Collection
77	The Segregation Model	93	Tracer Results
77	The Dispersion Model	93	Data Analysis
79	The Tanks-in-Series Model	93	Data Interpretation
80	Tracer Data Interpretation	93	Other Criteria That Influence
80	The *d* and *N* Indices		Disinfection Efficiencies
80	Other Index Values	97	Reactor Design Considerations
82	Tracer Analysis Examples	97	Baffles
82	Example 1: Step-Feed Test	97	Configuration
82	Data Collection	97	Surrounding Conditions
82	Tracer Results	98	References
83	Data Analysis	99	Suggested Readings
84	Data Interpretation		

INTRODUCTION

Wastewater disinfection efficiency may be influenced by a number of variables from the following three general categories:

- Wastewater and other surrounding conditions (chemical/physical),
- Disinfectant properties (kinetics), and
- Hydraulic characteristics of the reactor vessel.

As with most municipal wastewater treatment processes, the first category is difficult or impossible to control because it is often a function of the influent and other natural characteristics. Differences in disinfection efficiencies for various treatment effluents, such as primary or secondary discharges, are, however, important considerations and are covered in other sections of this manual. The second category primarily depends on the type of disinfectant selected. Disinfection agents will vary in their reaction rates, decay reactions, and mechanisms of biological deactivation, and will therefore influence the efficiency of disinfection as well as the unit process design. These relationships are also covered elsewhere. This chapter focuses on the hydraulic characteristics of a reactor vessel (third category) and shows how these characteristics can influence disinfection efficiency and therefore should be considered at the design and operating level.

A reactor's hydraulic characteristics are identified through tracer experiments. Fundamentals of tracer data analysis and data generation are presented in this chapter, along with step-by-step examples for analyzing these data.

The objectives of this chapter are to

- Outline fundamental concepts of reactor dynamics,
- Describe tracer testing methods,
- Describe and illustrate tracer data and interpretation methods, and
- Illustrate the use of tracer curves plus disinfection kinetic equations to predict a disinfection reactor's efficiency.

IDEAL REACTORS

Hydraulic characteristics of disinfection units may be thought of as variations of two ideal types of reactors:

- Plug flow, and
- Completely mixed.

The plug-flow reactor (PFR) operates under the assumption that all flow is unidirectional, with no mixing in the axial direction. An element of water or

---

influent entering a PFR (Figure 4.1) will travel from the inlet to the outlet for a period of time equal to the reactor volume (V) divided by the flow rate (Q). Any input stimuli (such as a tracer) injected at the inlet will exit in exactly the same manner after a lag time defined as V/Q, which is also the theoretical detention time.

Plug Flow

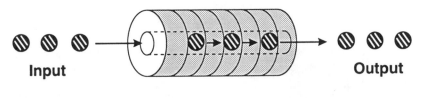

$$T = V/Q$$

Figure 4.1   Plug-flow reactor

A completely mixed reactor, also referred to as a *continuously stirred tank reactor* (CSTR), operates under the assumption that mixing is complete and instantaneous. An element of water or influent entering a CSTR (Figure 4.2), therefore, will become uniformly dispersed with all other elements of the water in the tank. If a stimuli slug were injected at the inlet, the stimuli concentration at the reactor outlet would initially equal the total mass of tracer divided by the reactor volume and then decrease at an exponential rate.

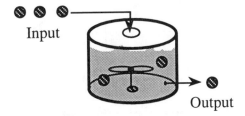

Figure 4.2   Continuously stirred tank reactor

The two tracer injection methods typically used for tracer experiments are the slug and step feed. These methods are discussed in the section Tracer Injection and Detection. The selected injection method will influence the resulting tracer output curve. Output curves similar to those in Figures 4.3 (a and b) and 4.4 (a and b) are known as the F and E curves and are discussed in the section Tracer Curves. Figures 4.3a and 4.4a illustrate the pattern of an ideal PFR and an ideal CSTR when a step input is used, while Figures 4.3b and 4.4b illustrate patterns when a pulse input is used. Methods used to generate these types of curves from tracer experiments are discussed in the next section.

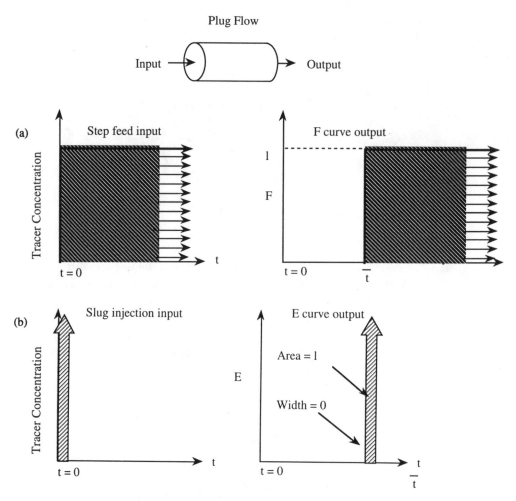

Figure 4.3   **Plug-flow reactor showing (a) step-feed input and *F* curve output and (b) slug-injection input and *E* curve output**

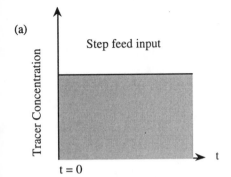
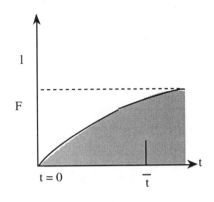

(a) Step feed input

Tracer Concentration

t = 0

1

F

t = 0 $\bar{t}$

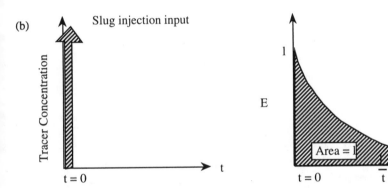

(b) Slug injection input

Tracer Concentration

t = 0

1

E

Area = 1

t = 0 $\bar{t}$

Figure 4.4   **Continuously stirred tank reactor showing (a) step-feed input and *F* curve output and (b) slug-injection input and *E* curve output**

T*RACER TEST METHODS*

Important areas of consideration when planning a tracer experiment are

- What flow conditions should be tested?
- What type of tracer should be used?
- How should the tracer be injected and monitored?

WASTEWATER FLOW CONDITIONS. A reactor's hydraulic character-istics and, subsequently, the resulting tracer curves and index calculations may respond differently under different hydraulic loadings. For this reason, it is important that tracer tests be conducted over a range of hydraulic loadings.

At a minimum, a reactor's hydraulic performance should be measured at conditions of peak flow and average daily flow. A peak-flow performance rating is important because disinfection units are typically sized according to worst-case conditions. Tracer testing at an average daily flow is also valuable, however, because these experiments will be useful in predicting the normal level of performance. In addition, experimental procedures, and consequently data accuracy, may be more reliable at lower than peak-flow conditions because adequate time will be available for gathering sufficient data. Flow conditions, and consequently the reactor's stability, should be as close to a steady-state condition as possible.

Unlike pilot-scale disinfection units, a full-scale operation has little or no control over the incoming flow rate. Operational records, therefore, must be consulted to identify times when desired flow conditions are to be expected and times when these flows should remain fairly constant. Disinfection units with relatively long residence times (such as chlorine contact chambers) are particularly susceptible to changes in hydraulic flow conditions during the tracer experiment.

TRACER SELECTION. Any acceptable material that does not disturb the flow pattern in the vessel can be used as a tracer. Various tracers such as dyes, salts, radioactive salts, and fluoride have been used. Selecting the proper tracer often involves an exercise in compromise. Obviously, the most desirable tracer would be an inert material that

- Does not react (chemically and/or physically) with the wastewater or the wastewater process or operation (conservative),
- Is easy to inject and measure,
- Is inexpensive and easy to handle,
- Is not normally found in background concentrations or is at least constant in the wastewater stream, and
- Will not contribute to a discharge violation.

Typically, one material that will satisfy most of these requirements, with the most notable exception of the first requirement, is an existing chemical used in the wastewater treatment process. Because a process chemical by nature may react with the wastewater, using such a chemical for the tracer agent should be approached with caution. The hydraulic behavior should influence how the tracer exits the reactor, not possible chemical and physical reactions. Table 4.1 lists some of the most commonly used tracers. Because each facility will have different characteristics, the final selection of an appropriate tracer may vary greatly. A careful observation of the above-listed requirements should be made before arriving at a selection.

Table 4.1   Typically used tracers

Tracer	Chemical	Analysis method	Comments	Reference
Fluoride	H_2SiF_6[a]	Ion electrode	May sorb onto solids	Teefy and Singer, 1990 VandeVenter et al., 1992
Sodium	NaCl[b]	Atomic absorption	Present in wastewater	Kreft et al., 1986 Teefy and Singer, 1990
Lithium	LiCl[c]	Atomic absorption	High analysis costs	Kreft et al., 1986 Nieuwstad et al., 1991
Dye	Rhodamine WT	Fluorometer	Reacts with oxidants	Severin, 1980
Chloride	NaCl	Mercuric nitrate method	Present in wastewater	—

[a] Fluosilicic acid.
[b] Sodium chloride.
[c] Lithium chloride.

TRACER INJECTION AND DETECTION. Two methods of tracer injection are pulse input and step input. Because both procedures can be used to generate the same type of data, the method selected is primarily dictated by properties of the reactor and other surrounding conditions. Obviously, the most important goal of the tracer experiment is to generate useful data; therefore, it is critical to make a sufficient number of accurate stimuli measurements. Methods of analyzing the data to generate performance indices will be most accurate if the stimuli are detected at regular time intervals. Tracer detection that can be continuously monitored, such as conductivity or fluorescence, is thus most desirable. The sensitivity and accuracy of the stimuli-measuring equipment should be set high enough to gauge a near to full recovery of the tracer and prevent any background interference from affecting the detection of low stimuli concentrations. In addition, the instrument's detection response time is important when short residence times are involved, such as with ultraviolet (UV) reactors.

Practical constraints, such as the ease of injecting and detecting the tracer, are always important considerations in selecting the right input method. Two critical factors regarding a successful tracer-injection operation are ensuring that successful premixing has occurred and that all tracer, or at least a known percentage of the tracer, enters the unit in question. Proper premixing may be achieved by injecting at a sufficient upstream location or through dynamic or static mixing. For cases where the tracer is injected upstream of parallel treatment units, one must be able to determine the flow distribution (and, consequently, the mass of tracer distribution) between those units. Making such an estimate is, of course, fairly risky.

A pulse input is accomplished by quickly injecting the tracer at the influent stream. The duration of the injection must be small relative to the reactor's hydraulic residence time. The amount injected must also be large enough to allow for ease of detection, but not so large that the outlet concentration is greater than any discharge permit requirements or beyond the detection range. The amount of tracer needed for a pulse injection, therefore, is not high, but the need to inject a slug quickly into the influent stream may cause some problems. The precise amount of tracer needed is difficult to predict without conducting some preliminary experiments (such as trial tracer injection tests). A first estimate can be made by setting the mass of tracer added, divided by the reactor volume, to a concentration that is slightly above the detection range. For a reactor with high short-circuiting properties or dead-space zones, this estimate would most likely be too high. In general, a disinfection unit with a long residence time (such as a chlorine contact chamber) will favor a pulse-injection approach, while a reactor with a short residence time (such as a UV unit) will not favor a pulse-injection approach. A primary advantage of this injection method is that a relatively small amount of tracer injection is needed, while a primary disadvantage is that the tracer peak may not be recorded unless a continuous monitoring method is used.

A step input is accomplished by feeding the tracer at the inlet until a steady-state condition is achieved at the effluent. Again, the concentration of this tracer feed will depend on reactor characteristics and detection sensitivity. The feed is then abruptly halted, and the resulting die-off is monitored at the discharge end until the background discharge concentration is reached. A step injection may also be achieved by starting with the background condition and initiating injection at time zero. Tracer feed is then continued until a steady-state concentration is reached at the discharge end. The duration of the injection, and consequently the amount of tracer used, for this procedure is higher than with a pulse-injection method. In general, a disinfection unit with a short residence time (UV unit) will favor a step-feed injection. A step-input approach is often used when a process feed chemical is used for the tracer. If a process feed chemical is not used, a primary disadvantage of this injection method is the necessity for a continuous tracer-feed setup and high tracer chemical costs.

As should be noted from the above discussions, a number of interrelated factors will influence a tracer experiment setup. A single recommended method, therefore, cannot be made. A summary of reported methods, such as presented in Table 4.1, should be helpful, however, in arriving at decisions necessary to complete a successful experimental design. The reader is therefore encouraged to review these and other references.

*T*RACER CURVES

Wastewater disinfection units will not perform exactly like an ideal PFR or complete-mix reactor; rather, it will produce tracer outputs that deviate from the typical response curves shown in Figures 4.3 and 4.4. Methods to graphically and mathematically describe these tracer outputs are presented below.

Confusion often results when using response curves (called *E*, *C*, and *F* curves) because the *x* axis is sometimes set as a time value (for example, minutes or seconds), and other times are set as ratios to the mean detention time (for example, 0.5, 1, and 1.5). This confusion becomes particularly relevant when index calculations are made. Other publications may use the nomenclature *E*(*t*) or *E*(Θ) to distinguish between plots that use actual time or relative time values for the *x* axis. All curves and index calculations presented here are based on an *x* axis that uses time values.

E **CURVE.** The residence time distribution (RTD) is based on an assumption that each element of water or influent traveling through a reactor will take a different route and, therefore, have a different residence time. The RTD, sometimes referred to as the distribution of ages of a reactor, may be defined as a function, *E*, such that *E* – *dt* is the fraction of material in the exiting stream with an age between *t* and *t* + *dt* (Levenspiel, 1972). Under ideal conditions, when there are no losses and no residual storage within the volume of the reactor, the *E* curve will represent the time taken for all of the injected elements to exit the system.

The *E* curve (Equation 4.1) may be developed from a pulse-input tracer test by dividing each effluent concentration at time t_i by the area under the concentration–time curve.

$$E = \frac{c_i}{\displaystyle\sum_{i=0}^{t_{total}} (c_i \Delta t_i)} \qquad (4.1)$$

Where

E	=	value at *y* axis of *C* curve, 1/time;
c_i	=	stimuli discharge concentration at time t_i, mass/volume;
t_{total}	=	total time of recovery; and
Δt_i	=	calculated difference between two time intervals on an *E*, *F*, or *C* curve.

The effluent concentration c_i is the midpoint of the two measured effluent concentration sample points. For better precision, this concentration is used instead of just the measured effluent concentration. Figure 4.5 shows the rela-

---

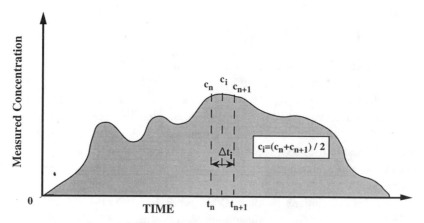

Figure 4.5   Calculating c_i using the midpoint method

tionship between the measured concentration (c_n and c_{n+1}), the measured times (t_n and t_{n+1}), and the calculated values of c_i and Δt_i.

The value t_i is any time between 0 and ∞ along the x axis under the E curve. The area bound by the E curve from 0 to t_i is the exit stream characterized by the fraction younger than age t_i and is defined as

$$\text{Area younger than } t_i = \int_0^{t_i} E \, dt \qquad (4.2)$$

Where

$t_i \quad = \quad$ time corresponding to the concentration c_i.

The E curve shown in Figure 4.6 illustrates that the area under this curve is unity. In other words,

$$\int_0^\infty E \, dt = 1 \qquad (4.3)$$

Figure 4.6 also identifies points of interest along the E curve.

The fraction older than time t_i may now be defined as

$$\text{Area older than } t_i = \int_{t_i}^\infty E \, dt = 1 - \int_0^{t_i} E \, dt \qquad (4.4)$$

Equations 4.2 and 4.4 are used later to develop relationships with the F curve.

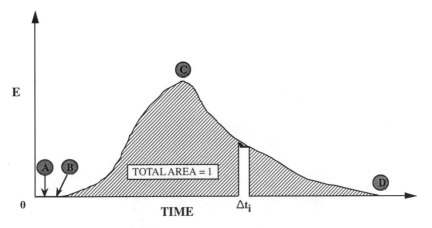

Figure 4.6   The *E* curve (A = no tracer detected [zero background];
B = first tracer element exits reactor; C = peak tracer
concentration; and D = last detected tracer element)

C **CURVE.** Many tracer studies report results with a *C* curve. Like the *E*
curve, the *C* curve is a plot of tracer output from a reactor versus time after a
pulse injection. A *C* curve is generated by dividing each effluent concentration c_i by the area under the concentration–time curve, $\sum (c_i \Delta t_i)$.

$$C = \frac{c_i}{\displaystyle\sum_{i=0}^{t_{\text{total}}} (c_i \Delta t_i)} \qquad (4.5)$$

As can be seen, the *C* curve is equivalent to the *E* curve.

$$C = E \qquad (4.6)$$

Thus, tracer studies that report data in the form of *C* curves are the same as
tracer studies reporting data in the *E* curve format.

F **CURVE.** Although the *C* and *E* curves may be found directly from pulse-
injection experiments, some experimental conditions (such as reactors with
short retention times), do not favor a pulse-injection approach. A step-feed
tracer experiment is therefore used. Data generated by this type of experiment
are illustrated in Figure 4.7. This curve has also been called the *die-off* concentration curve. As shown, the calculated c_i and t_i values are determined in a
manner similar to Figure 4.5. The value of *F* is found through Equation 4.7
when a tracer die-off, step-feed method is used. The resulting *F* curve is
shown in Figure 4.8.

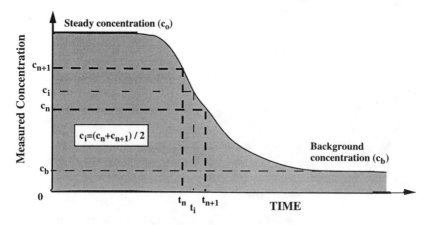

Figure 4.7   Measured concentrations to generate F curve

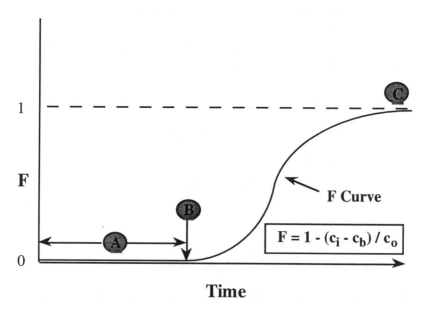

Figure 4.8   The F curve (A = tracer at steady state; B = input tracer feed shutoff; and C = all input tracer elements have exited system)

$$F = 1 - \frac{c_i - c_b}{c_o} \qquad (4.7)$$

Where

c_b = the background tracer concentration, mass/volume; and

c_o = the steady-state concentration, mass/volume.

A plot of the F curve ranges from 0 to 1. At point A, the input stimuli have reached steady state in the effluent stream. At the steady-state point, the input stimuli are shut off. This response is seen in the output effluent concentration at a time-lag point B. At point C, all elements of the stimuli input have exited the reactor. The fraction of tracer material exiting the vessel from 0 to t_i is equivalent to the fraction of material exiting the vessel that is younger than time t_i. This fraction has been previously defined in Equation 4.2. Thus, mathematically, the following is true:

$$F = \int_0^{t_i} E\,dt \qquad (4.8)$$

When both sides of Equation 4.8 are differentiated, a new equation arises:

$$dF/dT = E \qquad (4.9)$$

which is also equal to C.

Therefore, according to Equations 4.8 and 4.9, data from either a step-input or a pulse-tracer test can be used to develop an E or F curve. Index calculations of T_{mean} and σ^2 (to be discussed below) are therefore possible with either test method.

MEAN RESIDENCE TIME. From the C or E curve, the mean residence time T_{mean} can be determined by multiplying the sum of each time element by the area of that time element bounded by the C curve and dividing by the total area under the C curve. This is commonly known as the first moment of the RTD function (Fogler, 1993).

$$T_{mean} = \left(\int_0^{\infty} tC\,dt \right) \Big/ \left(\int_0^{\infty} C\,dt \right) \qquad (4.10)$$

In discrete measurements, T_{mean} can be algebraically determined as

$$T_{mean} = \left(\sum_{i=0}^{t_{total}} t_i c_i \Delta t_i \right) \Big/ \left(\sum_{i=0}^{t_{total}} c_i \Delta t_i \right) \qquad (4.11)$$

T_{mean} can be simplified to

$$T_{mean} = \left(\int_0^\infty tCdt\right)\bigg/\left(\int_0^\infty Cdt\right) = \left(\int_0^\infty tCdt\right)\bigg/(1)$$

$$= \left(\int_0^\infty tEdt\right) = \left(\sum t_i E_i \Delta t_i\right) \tag{4.12}$$

In other words $T_{mean} = T_{mean(C)} = T_{mean(E)}$.

For situations in which there is no loss in the effective volume of the reactor or no loss of tracer element as a result of leakage, the theoretical residence time, calculated as $T = V/Q$, is always equal to the mean residence time. Discrepancies of T_{mean} to the theoretical detention time can be used to judge the reliability of the tracer experiment (V calculations, tracer measurements, Q measurements) or the loss in reactor-effective volume (actual dead space). A T_{mean} less than T, therefore, may be interpreted as severe short-circuiting and/or dead space resulting from a loss in effective reactor volume.

VARIANCE. Variance σ^2 is a measure of the distribution spread and is mathematically described as

$$\sigma^2 = \left(\int_0^\infty (t - T_{mean})^2 Cdt\right)\bigg/\left(\int_0^\infty Cdt\right) \tag{4.13}$$

This is commonly known as the second moment about the mean (Fogler, 1993). Substituting Equation 4.10 into the expanded Equation 4.13 yields the following simplification:

$$\sigma^2 = \left[\left(\int_0^\infty (t)^2 Cdt\right)\bigg/\left(\int_0^\infty Cdt\right)\right] - (T_{mean})^2 \tag{4.14}$$

Discretely, in terms of concentration, the variance can be represented by the following:

$$\sigma^2 = \frac{\displaystyle\sum_{i=0}^{t_{total}} (t_i^2 c_i \Delta t_i)}{\displaystyle\sum_{i=0}^{t_{total}} c_i \Delta t_i} - T_{mean}^2 \tag{4.15}$$

From the E curve, variance may be calculated as

$$\sigma^2 = \sum_{i=0}^{t_{\text{total}}} (t_i^2 E_i \Delta t_i) - (T_{\text{mean}})^2 \qquad (4.16)$$

Figure 4.9 illustrates how the E or C curve shape will influence calculations of variance.

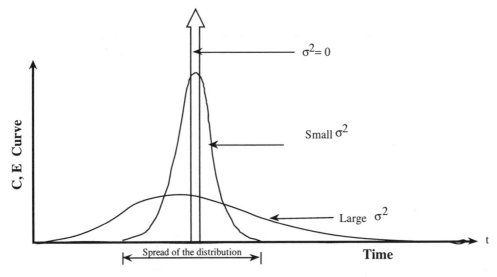

$\sigma^2 = 0$

Small σ^2

Large σ^2

Spread of the distribution

Time

t

C, E Curve

Figure 4.9   Effects of variance on the *C,E* curve

NONIDEAL REACTORS

In most situations, wastewater disinfection units do not behave as ideal PFRs or complete-mix reactors. Flow patterns in these reactors are a result of complex hydraulic patterns resulting from short-circuiting, back-mixing, dead-space stagnation, and turbulence. A description of typical wastewater disinfection units is presented below.

TYPICAL WASTEWATER DISINFECTION REACTORS. Serpentine flow reactors used for chlorine contact chambers and certain ozone contactors (see Figure 4.10) are designed to approach plug-flow conditions but may contain zones of back-mixing (particularly at the entrance zone), short-circuiting, and dead zones. This type of configuration has been found to be most successful in terms of approaching a plug-flow condition when length-to-width ratios of 10:1 and, preferably, 40:1 are used (Tchobanoglous and Burton, 1991). Modifications for existing units with lower length-to-width ratios have been suggested to improve the hydraulic pattern (Hart, 1979, and Louie and

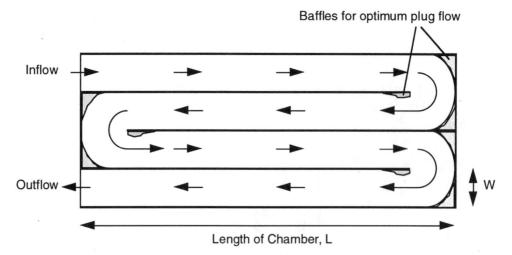

Figure 4.10 **Serpentine flow reactor (length of flow in this chamber is four times the length of the chamber) (from White, G.C. [1992]** *The Handbook of Chlorination and Alternative Disinfectants***. 3rd Ed., Van Nostrand Reinhold, New York, N.Y., with permission)**

Fohrman, 1968). Mixing conditions before the inlet of these units (especially for the chlorine and ozone disinfection processes) are considered by many to be beneficial because uniform distribution of the chemical will be ensured (Calmer, 1993).

Some ozone disinfection units are designed to approach a complete-mix condition (see Chapter 6). These units may also contain short-circuiting and dead-zone patterns that will result in a nonideal tracer response.

Ultraviolet disinfection units (see Figure 4.11), with their high length-to-width ratios, are designed to closely follow a plug-flow pattern. Inlet and outlet conditions for these reactors are important because of relative short detention times in the reactor units (White, 1992). Maximization of radial mixing (mixing perpendicular to flow) is a desirable feature of these disinfection units. This is unique to UV radiation reactors because radiation dose is proportional to the distance from the radiation source. Monitoring UV dose, therefore, is difficult.

CONCEPTUAL MODELS OF NONIDEAL REACTORS. Because actual reactors will not behave as ideal plug-flow or complete-mix models, other models must be adopted to explain their resulting RTD curves. Such a step is necessary to estimate the expected reactor disinfection efficiency.

This section discusses three additional models that are often used to describe the fluid flow pattern or RTD of nonideal reactors. It must be cautioned that these are conceptual models rather than absolute representations of the actual reactors. They do not necessarily have to parallel the disinfection reactor's configuration. As will be explained in a later section, the purpose for

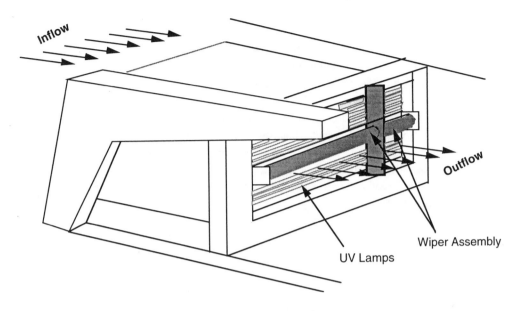

Figure 4.11 **Cross section of a typical ultraviolet reactor (from White, G.C. [1992]** *The Handbook of Chlorination and Alternative Disinfectants*. **3rd Ed., Van Nostrand Reinhold, New York, N.Y., with permission)**

adopting these conceptual reactor schemes is to predict a reactor's performance. For disinfection reactions, this means predicting the disinfection efficiency.

The Segregation Model. This model "visualizes" water elements flowing through a reactor as a mixture of individual batch reactors that will not exchange materials between them (hence the term *segregation*). These equally sized miniature batch reactors will remain in the nonideal reactor for a specific time before exiting. If the population of these miniature batch reactors were monitored at the discharge and plotted against time, the result would be an E curve. As will be illustrated later, a direct use of the RTD curve is therefore possible to estimate the reactor's efficiency. This is done by segmenting the E or C curve response and applying each individual segment to the expected disinfection reaction.

The Dispersion Model. The dispersion model operates under the assumption that flow through the reactor follows a plug-flow pattern but will deviate from the ideal plug-flow condition because of axial dispersion. A dimensionless number called the *axial dispersion number* (d) is defined as:

$$d = D/ul \qquad (4.17)$$

Where

D = the dispersion coefficient, units of length^2/time;

u = fluid velocity, length/time; and

l = length in the longitudinal direction of flow.

The inverse of d (ul/D) is termed the *Peclet number* (P_e). A dispersion number of 0 indicates that the reactor behaves as an ideal plug flow, while a dispersion number of infinity indicates that the reactor behaves as complete-mixed. No gross short-circuiting or dead-space conditions (such as severe channeling) are assumed to be present in this model. Figure 4.12 illustrates the relationship of D/ul to the E curve.

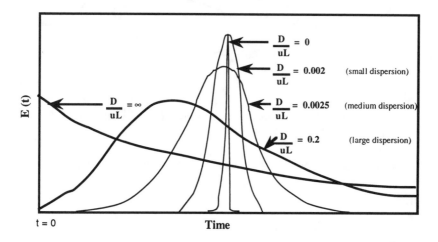

Figure 4.12   ***E* curve response for variations of the dispersion model (Levenspiel, 1972)**

The dispersion number may be estimated from mean residence time (T_{mean}) and variance (σ^2) values. For low dispersion (D/ul less than 0.01), D/ul can be approximated as follows:

$$\frac{\sigma^2}{\bar{t}^2} = \frac{\sigma^2}{(T_{mean})^2} = \frac{2D}{ul} \qquad (4.18)$$

Where

$\bar{t}^2$ = Square of mean residence time.

When Equation 4.18 yields higher dispersion values, other equations must be used depending on the condition of a closed reactor or an opened reactor. This definition of a reactor refers to inlet and outlet conditions. For a closed reactor, the dispersion at both inlet and outlet zones is considered minimal, while for an open reactor, the inlet and outlet are assumed to have high dispersion. Thus, the specific reactor itself, such as an ozone contactor or a UV reactor, is not the controlling factor in determining whether an open or closed reactor condition exists.

In the case of closed reactors for which Equation 4.18 is not applicable, the D/ul number may be approximated by as follows:

$$\frac{\sigma^2}{\bar{t}^2} = \frac{\sigma^2}{(T_{mean})^2} = 2\left(\frac{D}{ul}\right) - 2\left(\frac{D}{ul}\right)^2 (1 - e^{-ul/D}) \qquad (4.19)$$

For the use of open reactors where large dispersions occur, the D/ul number may be approximated as follows:

$$\frac{\sigma^2}{\bar{t}^2} = \frac{\sigma^2}{(T_{mean})^2} = 2\left(\frac{D}{ul}\right) + 8\left(\frac{D}{ul}\right)^2 \qquad (4.20)$$

The additional right-hand terms in Equations 4.19 and 4.20 will be negligible for small dispersion numbers. When large dispersion numbers are first approximated from Equation 4.18, a trial-and-error approach can be used with either Equation 4.19 or 4.20 to find D/ul. Equation 4.20 can also be solved using the quadratic formula.

The Tanks-in-Series Model. In the tanks-in-series model, the nonideal reactor is made up of a specific number of equally sized CSTR reactors (N), placed in a series configuration. Using T_{mean} and σ^2 parameters generated from these tracer curves, the value of N can also be calculated as follows:

$$N = \frac{\bar{t}^2}{\sigma^2} = \frac{(T_{mean})^2}{\sigma^2} \qquad (4.21)$$

Figure 4.13 illustrates how the E and F curves would appear for variations in the number of CSTR reactors. As shown, an N value of one varies according to one ideal CSTR, while a high N value approximates a PFR.

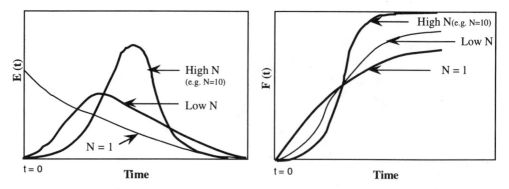

Figure 4.13   *E* and *F* curves for tanks-in-series model

T*RACER DATA INTERPRETATION*

Tracer data can either be interpreted directly or used to further predict the reactor's performance. Methods for predicting a reactor's disinfection efficiency through tracer curves and disinfection kinetic coefficients are presented later. This section discusses how one can interpret the results of tracer data by comparing that data to expected performance ratings.

THE *d* AND *N* INDICES. A dispersion number of 0.01 is commonly believed to represent a low dispersion condition, while a dispersion number between 0.01 and 0.1 represents a moderate dispersion condition, and a dispersion number greater than 0.1 indicates high dispersion. The work of Trussell and Chao (1977) suggests that improvements in process efficiency can be gained by lowering the dispersion number to approximately 0.01. Based on this information, many people have suggested that $d \approx 0.01$ is a reasonable target value for disinfection contact chambers. Similarly, an *N* value of 50 indicates a good PFR, an *N* value between 50 and 5 indicates a fair PFR, and an *N* value less than 5 indicates a poor PFR. Disinfection reactors that are designed to approach a plug-flow pattern should at least be rated with a *d* value between 0.01 and 0.1 or an *N* value between 50 and 5.

OTHER INDEX VALUES. In addition to *d* and *N*, a number of parameters based on the *E* or *C* curve response have been used to characterize the hydraulic performance of reactors. Although these indices are no longer believed to be useful and should be considered more qualitative than quantitative, they may still be reported in tracer studies. Thus, a description of these indices is in order (Figure 4.14 illustrates locations of important variables connected with these index values):

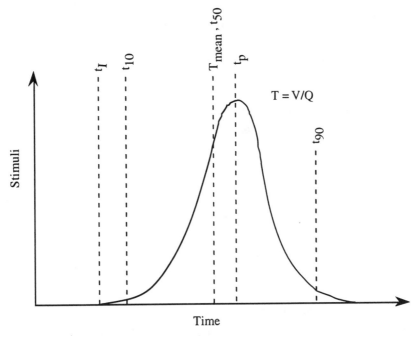

Figure 4.14 Typical trace curve parameters (in a closed system with no loss of effective volume, $T = T_{mean} = t_{50}$)

- The Morril dispersion index, t_{90}/t_{10}, is the ratio of the time required for 90% of the stimuli to disperse to the time required for 10% of the stimuli to disperse. A value of 1.0 would be obtained for an ideal plug-flow reactor, and a value of approximately 20 would be obtained for a completely mixed reactor (U.S. EPA, 1986).

- Index of average detention, T_{mean}/T, is the ratio of the mean residence time to the theoretical residence time (T = volume/flow rate). A value of 1.0 indicates an ideal plug-flow condition, while a value of less than 1.0 indicates either a loss of effective volume or short-circuiting.

- Index of mean detention, t_{50}/T, is the ratio of time required for 50% of the stimuli to disperse to the theoretical residence time. A value significantly less than 1.0 indicates a long tail in the E curve.

- Index of initial short-circuiting, t_I/T, is the ratio of time elapsed until stimuli is initially noted to the theoretical residence time. A value of 1.0 would be obtained in a plug-flow reactor, while a value significantly less than 1.0 indicates initial severe short-circuiting. The water industry currently bases disinfection efficiencies on a similar parameter (t_{10}/T) with the Surface Water Treatment rule.

- Index of modal detention time, t_p/T, is the ratio of time elapsed until a peak stimulus is noted to the theoretical residence time. A value of 1.0 would be obtained in a plug-flow reactor, while a number significantly less than 1.0 indicates short-circuiting or dead space.

In addition to calculating the index values listed above (including d and N), care should be taken to assess the reliability of a tracer experiment. A percent tracer recovery should be calculated and the reactor visually observed. A low tracer-recovery percentage indicates faulty experimental procedures such as inaccurate stimuli measurements, poor sampling locations, or errors in flow measurements. A 90 to 100% recovery is desirable. In addition, the value of T_{mean} should closely approximate the theoretical detention time of T. Visual observations may detect inappropriate sampling procedures, severe short-circuiting problems, and other possible difficulties. The value and need for visual observations should not be underestimated.

*T*RACER ANALYSIS EXAMPLES

This section presents two examples of tracer data analysis to illustrate how step-feed and pulse-feed tracer data may be used to calculate T_{mean} and σ^2 and, consequently, the N and d index numbers.

EXAMPLE 1: STEP-FEED TEST. Data Collection. A tracer dye was continuously fed to the inlet of a reactor until a steady-state discharge concentration of 5.0 mg/L was reached. Before initiating this tracer test, a background measurement of 0.25 mg/L was noted. The actual tracer concentration at steady state, therefore, was 4.75 mg/L. Throughout this tracer test, the influent flow rate was held steady at 473.1 ML/d (125 mgd). After steady state was reached, the tracer feed was shut off, and the resulting discharge concentration was recorded until the background concentration was reached.

A schematic of this tracer test and the raw data are shown in Figure 4.15.

Tracer Results. Figure 4.16 is a tabular output of data for generating the F curve (column 7), E curve (column 8), and percent tracer recovery (column 11). The T_{mean} value was found by summing column 9 (Equation 4.12), while σ^2 was found by subtracting the square of T_{mean} from the sum of column 10 (Equation 4.16).

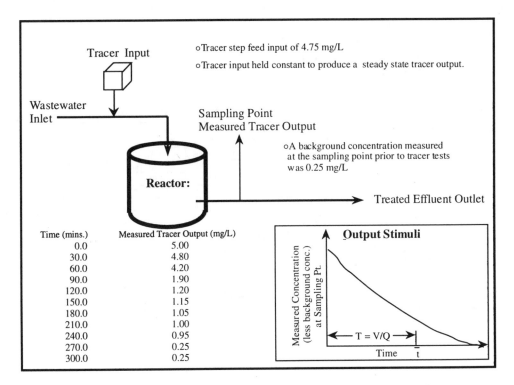

Figure 4.15   Example 1 (step feed)

Data Analysis. Resulting F and E curves are shown in Figures 4.17 and 4.18, respectively.

The dispersion index (d) is first estimated by Equation 4.18 to be

$$\frac{4\ 753}{(116.5)^2} = 2d$$

$$d = 0.18$$

Because this number is greater than 0.01, Equation 4.19 or 4.20 must be used. Through a trial-and-error approach, the dispersion number for an opened reactor (Equation 4.20) is found to be 0.12, while the dispersion number for a closed reactor (Equation 4.19) is found to be 0.23.

The value of N is found by Equation 4.21 to be

$$N = \frac{(116.5)^2}{4\ 753}$$

$$N = 2.86$$

I. INITIAL CONDITIONS

Reactor Volume in MG	=	*10*	37.85 M liters
Flow Rate into Reactor (Q) in MGD	=	*125*	473.13 M liters/D
Background Concentration in mg/L	=	*0.25*	
Tracer Concentration at Steady-State in mg/L	=	*4.75*	

II. DATA COLLECTION AND ANALYSIS

Time t (mins.)	dt	Measured Conc. mg/L	Actual Conc.	Average Conc. (c)	c/co	F-curve 1 - c/co	E-curve dF/dt	tEdt	t^2Edt	Recovery Qcdt
0.0		*5.00*	4.75	4.75	1.000	0.000	0.0000	0.00	0.00	0
30.0	30	*4.80*	4.55	4.65	0.979	0.021	0.0007	0.63	18.95	45773
60.0	30	*4.20*	3.95	4.25	0.895	0.105	0.0028	5.05	303.16	41836
90.0	30	*1.90*	1.65	2.80	0.589	0.411	0.0102	27.47	2472.63	27563
120.0	30	*1.20*	0.95	1.30	0.274	0.726	0.0105	37.89	4547.37	12797
150.0	30	*1.15*	0.90	0.93	0.195	0.805	0.0026	11.84	1776.32	9105
180.0	30	*1.05*	0.80	0.85	0.179	0.821	0.0005	2.84	511.58	8367
210.0	30	*1.00*	0.75	0.78	0.163	0.837	0.0005	3.32	696.32	7629
240.0	30	*0.95*	0.70	0.73	0.153	0.847	0.0004	2.53	606.32	7137
270.0	30	*0.90*	0.65	0.68	0.142	0.858	0.0004	2.84	767.37	6645
300.0	30	*0.25*	0.00	0.33	0.068	0.932	0.0025	22.11	6631.58	3199
330.0	30	*0.25*	0.00	0.00	0.000	1.000	0.0000	0.00	0.00	0
							SUMMATIONS =	116.53	18331.58	170051

III. TRACER RESULTS

Tracer Recovered in grams	=	170,051
Tracer Added in grams	=	179,550
Percent Recovered	=	95 %
T mean in mins.	=	116.5
Theoretical Detention Time in mins	=	115.2
Variance	=	4753

Figure 4.16   Example 1 data output

Data Interpretation. A tracer recovery of 95% and T_{mean} value of 117 minutes (which is close to the theoretical detention time of 115 minutes) indicates that this tracer experiment is acceptable.

The dispersion values and N value indicate that this reactor does not follow a good plug-flow pattern.

EXAMPLE 2: PULSE-FEED TEST. Data Collection. A single 2 500-g pulse (slug) of tracer material was injected into a reactor. Discharge tracer levels were recorded from the initial injection time (time = 0) to the completion of the experiment. Before initiating this tracer test, a background measurement of 5.5 mg/L was noted. Throughout the test, the flow rate into the reactor was held constant at 53.1 ML/d (13.7 mgd).

A schematic of this tracer test and the raw data are shown in Figure 4.19.

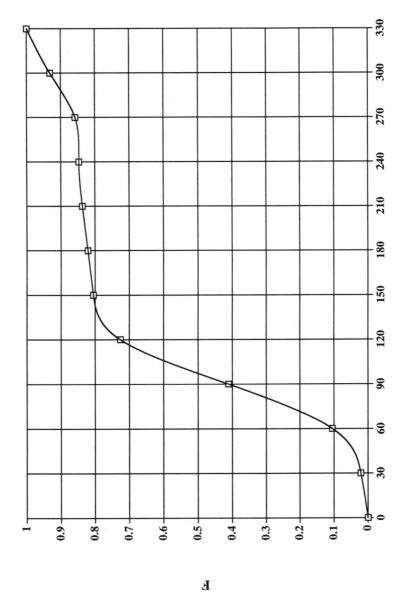

F

Figure 4.17   Example 1 (F curve)

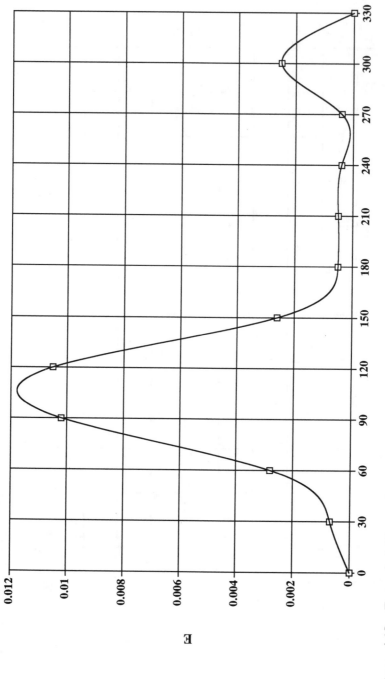

Figure 4.18 Example 1 (*E* curve)

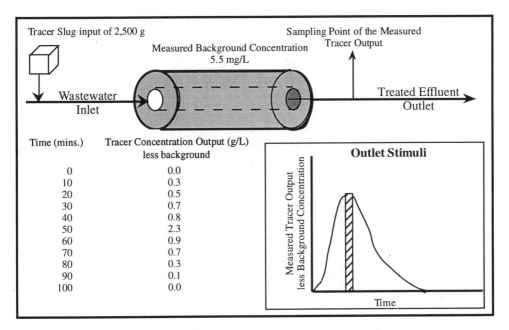

Time (mins.)	Tracer Concentration Output (g/L) less background
0	0.0
10	0.3
20	0.5
30	0.7
40	0.8
50	2.3
60	0.9
70	0.7
80	0.3
90	0.1
100	0.0

Figure 4.19   Example 2 (pulse feed)

Tracer Results. Figure 4.20 is a tabular output of data for generating the E curve (column 7) and the percent tracer recovery (column 10). The T_{mean} value was found by summing column 8 (Equation 4.12), and σ^2 was found by subtracting the square of T_{mean} from the sum of column 10 (Equation 4.16).

Data Analysis. The resulting E curve is shown in Figure 4.21. The dispersion index (d) may first be estimated using Equation 4.18 to be

$$2d = \frac{337.8}{(53.03)^2} = 0.12$$

$$d = 0.06$$

Because this number is greater than 0.01, Equation 4.19 or 4.20 must be used. Through trial and error, the dispersion number for an opened reactor (Equation 4.20) is found to be 0.05, while the dispersion number for a closed reactor (Equation 4.19) is found to be 0.06.

The value of N is found by Equation 4.21 to be

$$N = \frac{(53.03)^2}{337.8} = 8.3$$

Data Interpretation. A tracer recovery of 95% and T_{mean} value of 53 minutes (which is close to the theoretical detention time) indicates that data resulting from this tracer experiment are acceptable.

I.   **INITIAL CONDITIONS**

Volume of Reactor in MG	=	0.5	1.94 M liters
Flow into Reactor in MGD	=	13.7	53.1 M liters/D
Background Concentration in mg/L	=	5.50	
Amount of Tracer Added (g)	=	2500	

II.   **DATA COLLECTION AND ANALYSIS**

Time t (mins)	dt	Measured Conc. mg/l	Actual Conc.	Average c	c dt	E-curve (eq. 2)	tEdt	t^2Edt	Recovered Q c dt
0		5.50	0.00	0.000	0.000	0.000	0.00	0.00	0.00
10	10	5.80	0.30	0.150	1.500	0.002	0.23	2.27	53.94
20	10	6.00	0.50	0.400	4.000	0.006	1.21	24.24	143.85
30	10	6.20	0.70	0.600	6.000	0.009	2.73	81.82	215.78
40	10	6.30	0.80	0.750	7.500	0.011	4.55	181.82	269.72
50	10	7.80	2.30	1.550	15.500	0.023	11.74	587.12	557.42
60	10	6.40	0.90	1.600	16.000	0.024	14.55	872.73	575.40
70	10	6.20	0.70	0.800	8.000	0.012	8.48	593.94	287.70
80	10	5.80	0.30	0.500	5.000	0.008	6.06	484.85	179.81
90	10	5.60	0.10	0.200	2.000	0.003	2.73	245.45	71.92
100	10	5.50	0.00	0.050	0.500	0.001	0.76	75.76	17.98
SUMMATIONS =					66.000		53.030	3,150.000	2373.53

III.   **TRACER RESULTS**

Tracer Recovered in grams	=	2374
Tracer Added in grams	=	2500
Percent Recovered	=	95 %
T mean	=	53
Theoretical Residence Time	=	53
Variance	=	337.79

Figure 4.20   Example 2 data output

The dispersion index and N values indicate that this reactor is fair to good in terms of approaching a plug-flow pattern.

COMBINING TRACER ANALYSIS AND DISINFECTION KINETICS

A useful analytical procedure for predicting the efficiency of a wastewater disinfection reactor is to assume that the reactor behaves according to a simplified model (such as the ideal plug-flow or complete-mix model, the segregation model, the dispersion model, or the tanks-in-series model). By combining equations that describe these simplified models with equations that define disinfection kinetics, one can predict how that model would control the

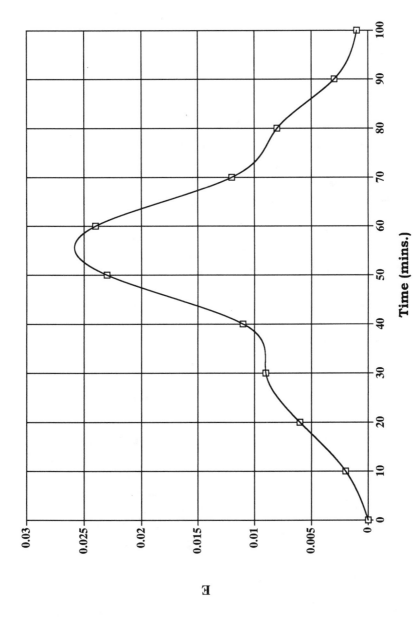

Figure 4.21   Example 2 (*E* curve)

disinfection process as wastewater passes through. The following section describes this concept and presents an example calculation.

DISINFECTION KINETICS. The most frequently used method for describing wastewater disinfection kinetics is the application of various forms of the Chick-Watson equation. This equation, first introduced in 1908 (Chick, 1908, and Watson, 1908), may be written as follows:

$$\frac{I}{I_o} = e^{-k'C_x t_i} \tag{4.22}$$

Where

I =   microorganism population at time t_i,
I_o =   microorganism population at time 0,
t_i =   time,
C_x =   disinfection residual, and
k' =   reaction rate (1/[residual × time]).

(Note that the nomenclature I and I_o is used in this chapter because N is the conventional variable used in reactor dynamics studies to signify the number of CSTRs. Conventional nomenclature for microorganism populations [as used in other chapters of this manual] is N and N_o.)

For chlorine and ozone, C_x may be expressed in milligrams per litre, while for UV radiation, C_x may be expressed in microwatts per square centimetre. Often, the k' and C_x terms are combined to form another first-order disinfection rate, k, in units of 1/time.

$$\frac{I}{I_o} = e^{-k t_i} \tag{4.23}$$

Equation 4.23 is, of course, a simplification of actual disinfection mechanisms, as many studies have reported deviations from the Chick-Watson equation. Adopting a first-order reaction rate, however, simplifies the reactor performance evaluations. All disinfection reaction k values used in equations presented below assume the reaction rate is first order.

These disinfection coefficients are sensitive to water conditions, types of microorganisms, and physical conditions. In addition, chlorine, chlorine dioxide, and ozone are unstable when applied to wastewater, making the assumption that C remains constant incorrect (Haas *et al.*, 1994). When precise reactor performance evaluations, rather than relative performance ratings, are needed, bench-scale kinetic tests and a consideration of models that include dynamic conditions are advisable to identify the disinfection kinetics.

IDEAL MODELS. In an ideal plug-flow reactor, the reaction time is equal to T_{mean} because a plug injection will exit the reactor in a single slug that occurs at T_{mean}. Equation 4.23 can therefore be used directly to estimate I/I_o of a plug-flow reactor simply by substituting T_{mean} for t_i. Because the lefthand term of Equation 4.23 is equal to the percentage of microorganisms remaining after disinfection (in decimal form), 1.0 minus that term will equal the disinfection efficiency as follows:

$$\text{Plug flow efficiency} = 1 - \frac{I}{I_o} = 1 - e^{-kt_i} \tag{4.24}$$

In many chemical engineering reactor texts, the efficiency is referred to as the conversion, or X, factor.

For an ideal CSTR, the E function may be written as follows (Fogler, 1993):

$$E = \frac{1}{T_{\text{mean}}} e^{-t_i/T_{\text{mean}}} \tag{4.25}$$

The overall disinfection efficiency for this type of reactor must account for mixing and, thus, can be evaluated as

$$1 - \int_0^\infty e^{-kt_i} E \, dt \tag{4.26}$$

Substituting Equation 4.25 into Equation 4.26 yields the following:

$$\text{Complete mix efficiency} = \frac{T_{\text{mean}} \, k}{1 + T_{\text{mean}} \, k} \tag{4.27}$$

Disinfection efficiencies for ideal reactors, therefore, may be estimated through Equation 4.24 for plug-flow or Equation 4.27 for complete-mix reactors.

USING THE SEGREGATED MODEL. The segregated model operates on the assumption that flow through the disinfection unit consists of individual groups or elements that maintain their own identities. The relative number of elements passing out of the reactor at any one time will equal the value of E. Because these elements maintain their own identities, they may be thought of as miniature batch reactors with relative sizes equal to the value of E and reaction times equal to t_i. The overall efficiency of the disinfection reactor, therefore, may be determined as follows (Fogler, 1993):

$$\text{Efficiency} = \sum_{0}^{\infty}(1 - e^{-kt_i})E_i\Delta t_i \qquad (4.28)$$

USING THE TANKS-IN-SERIES MODEL. The tanks-in-series model treats the disinfection reactor as consisting of a series of equally sized CSTRs; the output of an upstream reactor equals the input to the downstream reactor. An equation for estimating the overall efficiency of this reactor series is as follows (Fogler, 1993):

$$\text{Efficiency} = 1 - \frac{1}{[1 + T_{\text{mean}}/Nk^N]} \qquad (4.29)$$

USING THE DISPERSION MODEL. The dispersion model operates on the assumption that flow passing through a reactor is influenced by dispersion. For example, if a pulse injection were placed at the inlet of a long tubular reactor, that pulse spike would spread out as it passed down the length of flow. The degree of spread would depend on the amount of dispersion occurring in the reactor. Figure 4.12 illustrated this effect on dispersion. An equation for estimating the overall efficiency of this reactor is as follows (Fogler, 1993):

$$\text{Efficiency} = 1 - \frac{4qe^{\frac{1}{2d}}}{(1+q)^2 e^{\frac{q}{2d}} - (1-q)^2 e^{-\frac{q}{2d}}} \qquad (4.30)$$

Where

$$q = (1 + 4k\tau d)^{\frac{1}{2}},$$

$$\tau = T_{\text{mean}} \text{ (for closed vessels), and}$$

$$\tau = \frac{T_{\text{mean}}}{1 + 2d} \text{ (for open vessels).}$$

Example calculations of disinfection efficiencies using Equations 4.24, 4.27, 4.28, 4.29, and 4.30 are presented below.

CALCULATING A REACTOR'S POTENTIAL EFFICIENCY

As noted in the tracer analysis examples, a reactor's performance can be estimated strictly through tracer curves and the resulting index calculations. A more informative estimate of a reactor's performance capability, however, would be to also consider disinfection kinetics as presented in the preceding section.

The example problem in this section illustrates how one can combine tracer results with disinfection kinetics to rate the overall performance of a disinfection reactor.

EXAMPLE 3: DISINFECTION EFFICIENCY ESTIMATION. Data Collection. The tracer test conducted for this example is similar to the Example 2 test above. Conditions of this test are summarized as follows:

- Tracer injection: pulse, 1 350 g;
- Background concentration: 0.0 mg/L;
- Reactor volume: 153.5 L (39.6 gal); and
- Flow rate: 38 750 L/d (10 000 gpd).

Separate disinfection kinetic studies of this wastewater and the disinfectant in question resulted in a k value of 0.05 (residual min)$^{-1}$. At a dose of 5.00 mg/L, this resulted in a k' coefficient of 0.25 min^{-1}.

Tracer Results. Figure 4.22 is a tabular output of the tracer data and calculations of T_{mean} and σ^2. Methods for calculating these indices are the same as those in Example 2.

Data Analysis. The resulting E curve for this tracer test is shown in Figure 4.23. Using the same methods as described for Example 2, the dispersion index d was found to be 0.08 for an opened reactor and 0.12 for a closed reactor. The N number was calculated to be 4.73.

Further disinfection performance estimates of this reactor are presented in Figure 4.24. As can be seen, the reactor's disinfection performance rating varied from 75.8% for an ideal PFR to 58.7 for a single CSTR.

Data Interpretation. A percent tracer recovery of 98% and T_{mean} of 5.7 minutes (which is close to the theoretical detention time) indicates that the data generated from this tracer experiment are acceptable. Depending on the type of reactor model selected, disinfection efficiencies will vary. Most models indicated that this reactor should perform at approximately a 70% efficiency (see Figure 4.24).

OTHER CRITERIA THAT INFLUENCE DISINFECTION EFFICIENCIES. The example presented above should not be interpreted as an exact solution of disinfection efficiency. The focus of this chapter has been to discuss hydraulic characteristics revealed through tracer analyses that are detected according to the residence time distribution. The means by which hydraulic characteristics influence the subsequent disinfection efficiency, as

INITIAL CONDITIONS

Volume of Reactor	=	39.6	Gallons or	153.5 liters
Flow into Reactor	=	0.01	MGD or	38750 liters/day
Background Conc.	=	0.00	g/L	
Amount of Tracer Added	=	1350	g NaCl	

DATA COLLECTION AND ANALYSIS

Time, T in mins.	dT	Measured Conc. g/L	Actual Conc.	Average c	c dt	E Curve (eq. 2)	t Edt	t^2Edt	Na Recovery Q c dt
0.00		0.0	0.00	0.000	0.000	0.000	0.000	0.000	0.00
1.00	1.00	1.0	1.00	0.500	0.500	0.010	0.01	0.010	13.13
2.00	1.00	5.0	5.00	3.000	3.000	0.059	0.12	0.237	78.75
3.00	1.00	8.0	8.00	6.500	6.500	0.128	0.38	1.155	170.63
4.00	1.00	10.0	10.00	9.000	9.000	0.178	0.71	2.843	236.25
5.00	1.00	8.0	8.00	9.000	9.000	0.178	0.89	4.442	236.25
6.00	1.00	6.0	6.00	7.000	7.000	0.138	0.83	4.975	183.75
7.00	1.00	4.0	4.00	5.000	5.000	0.099	0.69	4.837	131.25
8.00	1.00	3.0	3.00	3.500	3.500	0.069	0.55	4.423	91.88
9.00	1.00	2.2	2.20	2.600	2.600	0.051	0.46	4.158	68.25
10.00	1.00	1.5	1.50	1.850	1.850	0.037	0.37	3.653	48.56
12.00	2.00	0.6	0.60	1.050	2.100	0.021	0.50	5.970	55.13
14.00	2.00	0.0	0.00	0.300	0.600	0.006	0.17	2.322	15.75
					49.3	50.65		5.68	39.02	1329.56

TRACER RESULTS

Total Recovered in grams	=	1329.56
Total Added in grams	=	1350.00
Percent Recovered	=	98%
T mean in mins.	=	5.7
Theoretical Detention Time in mins.	=	5.7
Variance	=	6.81

Figure 4.22   Example 3 data output

presented above, emphasize the significance of hydraulics when evaluating and designing a disinfection reactor. Clearly, a number of other parameters can have an effect on a reactor's performance. This section and other chapters found in this manual identify some of these parameters for chlorine contact chambers, ozone contactors, and UV reactors.

Under dynamic wastewater conditions, the chlorine residual will certainly decay. Thus, reactors with long residence times (such as a chlorine contact chamber) will not perform simply according to the Chick-Watson equation, which assumes that C_x is constant, but will be influenced by changes in the C_x value. The findings of those who have researched this area are discussed in Chapter 6.

Also, the concept of ozone transfer efficiency, as discussed in Chapter 6, is important because it describes how to maximize the contacting of ozone, a gas-liquid, to wastewater. Rakness *et al.* (1984) recommend that the designer require an 85% or better transfer efficiency.

In UV reactors, a designer tries to maximize the illumination efficiency. Laws of optical physics, which are important phenomena, are covered in

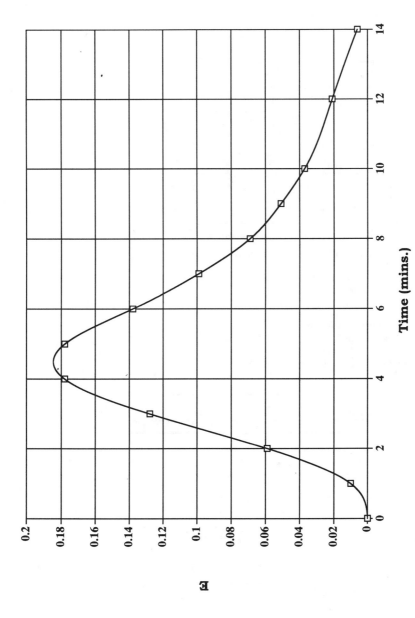

Figure 4.23   Example 3 (*E* curve)

IV. INITIAL PARAMETERS

Disinfection First Order Kinetics

Disinfection Dose (Cx)	=	5.00 mg/L
1st order Reaction Rate (k)	=	0.05
k' which is (k)(Cx)	=	0.25 min^-1

V. EFFICIENCY OF IDEAL REACTORS

Ideal PFR: 1-EXP((Tmean)*(k')*(-1))	=	0.758	or	75.8%
Ideal CSTR: (Tmean*k')/(1+(Tmean*k'))	=	0.587	or	58.7%

VI. EFFICIENCY OF SEGREGATED MODEL

Time, T in mins.	E(t)	X(t)	E(t) X(t) (1/min)
0.00	0.000	0.000	0.000
1.00	0.010	0.221	0.002
2.00	0.059	0.393	0.023
3.00	0.128	0.528	0.068
4.00	0.178	0.632	0.112
5.00	0.178	0.713	0.127
6.00	0.138	0.777	0.107
7.00	0.099	0.826	0.082
8.00	0.069	0.865	0.060
9.00	0.051	0.895	0.046
10.00	0.037	0.918	0.034
12.00	0.021	0.950	0.020
14.00	0.006	0.970	0.006
			0.686

Segregated Model: sum of (E(t) X(t)) = 0.686 or 68.6%

VII. EFFICIENCY OF TANK-IN-SERIES

N Tank in Series: Tmean^2/variance	=	4.73		
Efficiency: 1-(1/(1+(Tmean/N)*k')^N)	=	0.711	or	71.1%

VIII. DISPERSION NUMBER CALCULATIONS

Initial Estimates of (d) = 0.11

since d > 0.01 use below for approximation of d

Open Vessel	(variance/((t-mean^2)=2*d-8*d^2
Closed Vessel	(variance/(Tmean^2)=((2d^2)*(1/d - 1 + exp(-1/d)))

Approximating d for	Right	Left	Peclet No.	Vary D
Open dispersion vessel	0.211	0.211	11.40	0.08
Closed dispersion vessel	0.211	0.211	12.50	0.12

Right refers to ride side of the vessel dispersion equation.
Left refers to left side of the vessel dispersionequation.
The user must estimate a "d" value to balance the equation.

Efficiency of Closed Dispersion Vessel: (eq. 29)	=	0.714	or	71.4%
Efficiency of Open Dispersion Vessel: (eq. 29)	=	0.677	or	67.7%

IX.   SUMMARY OF DISINFECTION EFFICIENCIES

Ideal PFR	0.758	75.8%
Single CSTR	0.587	58.7%
Segregated Model	0.686	68.6%
Tank-in-Series	0.711	71.1%
Dispersion Closed	0.714	71.4%
Dispersion Open	0.677	67.7%

Figure 4.24   Example 3 disinfection analysis

Chapter 7. The *Municipal Wastewater Disinfection Manual* (U.S. EPA, 1986) also covers this topic.

REACTOR DESIGN CONSIDERATIONS

Although tracer analyses of disinfection reactors are typically conducted after a reactor is constructed, a knowledge of reactor dynamics could and should be used as a design tool. Throughout this chapter, desirable hydraulic features of a disinfection reactor have been noted. This section reviews some properties of a disinfection reactor that should be specified during the design process.

BAFFLES. Marske and Boyle (1973) proved that longitudinal baffled serpentine reactors or special annual ring clarifiers performed better than circular chambers. In their field evaluations of several chlorine contact chambers, they found that longitudinal baffled chambers with a flow length-to-width ratio of 72:1 can provide a 95% plug flow (see Figure 4.10). Longitudinal baffles were also found to be superior to horizontal baffles, which cause more back-mixing. Others have also investigated modifications of baffles to improve residence time distributions and dispersion indices (Hart, 1979, and Tchobanoglous and Burton, 1991). Details of baffle design may be found in Chapter 5.

CONFIGURATION. Another factor that should be considered by the designer is the depth-to-width (H/W) ratio. From the works of Calmer and Adams (1977), Sepp and Bao (1980), and Trussell and Chao (1977), the H/W ratio should be 1.0 or less.

Contact chambers should also have smooth corners to minimize dead space and short-circuiting. A smooth surface will also guard against the potential of bacterial growth within protected grooves. Suspended solids can be easily flushed out of the chamber during cleaning. Most important, many evaluations of various designs have reaffirmed that long narrow channels with longitudinal baffles make the best contact chambers. This chamber attribute most likely influences the plug-flow nature of the reactor and the need for low D/uL axial velocities. White (1992) details some of these design properties.

SURROUNDING CONDITIONS. The importance of initial mixing in chlorination and ozonation is also noted in Chapters 5 and 6. This may be accomplished at the inlet zone using both static and mechanical mixers.

In open-channel reactors, the designer must consider wind effects that may cause surface currents and, consequently, short-circuiting and the disruption of an acceptable plug-flow condition. When monochloramine is used as the disinfection agent, turbulence within the chamber should be minimized because it may cause back-mixing and reduce the concentration of volatile monochloramine (White, 1992).

*R*EFERENCES

Calmer, J.C. (1993) Chlorine Mixing Energy Requirements for Disinfection of Municipal Effluents. *Proc. Water Environ. Fed. Disinfect. Spec. Conf.*, Whippany, N.J.

Calmer, J.C., and Adams, R.M. (1977) *Design Guide Chlorination-Dechlorination Contact Facilities*. Kennedy/Jenks Engineers, San Francisco, Calif.

Chick, H. (1908) An Investigation of the Laws of Disinfection. *J. Hyg. (G.B.)*, **8**, 1908.

Fogler, H.S. (1993) *Elements of Chemical Reaction Engineering*. Prentice Hall, Englewood Cliffs, N.J.

Haas, C.N., *et al.* (1994) *Development and Validation of Rational Design Methods of Disinfection*. Am. Water Works Assoc. and Am. Water Works Res. Found., Denver, Colo.

Hart, F. (1979) Improved Hydraulic Performance of Chlorine Contact Chambers. *J. Water Pollut. Control Fed.*, **51**, 2868.

Kreft, P., *et al.* (1986) Hydraulic Studies and Cleaning Evaluations of Ultraviolet Disinfection Units. *J. Water Pollut. Control Fed.*, **51**, 2868.

Levenspiel, O. (1972) *Chemical Reaction Engineering*. John Wiley & Sons, New York, N.Y.

Louie, D., and Fohrman, M. (1968) Hydraulic Model Studies of Chlorine Mixing and Contact Chambers. *J. Water Pollut. Control Fed.*, **40**, 2.

Marske, D.M., and Boyle, V.D. (1973) Chlorine Contact Chamber Design—A Field Evaluation. *Water Sew. Works*, **120**, 70.

Nieuwstad, T., *et al.* (1991) Hydraulic and Microbiological Characterization of Reactors for Ultraviolet Disinfection of Secondary Wastewater Effluents. *J. Water Resour. Plann. Manage. Div., Proc. Am. Soc. Civ. Eng.*, **25**, 775.

Rakness, K.L., *et al.* (1984) Design, Start-Up, and Operation of an Ozone Disinfection Unit. *J. Water Pollut. Control Fed.*, **56**, 1152.

Sepp, E., and Bao, P. (1980) *Design Optimization of the Chlorination Process. Vol. I, Comparison of Optimized Pilot System With Existing Full-Scale Systems*. Wastewater Res., Munic. Environ. Res. Lab.

Severin (1980) Disinfection of Municipal Effluents with Ultraviolet Light. *J. Water Pollut. Control Fed.*, **52**, 2007.

Tchobanoglous, G., and Burton, F.L. (1991) *Wastewater Engineering*. 3rd Ed., McGraw-Hill, Inc., New York, N.Y.

Teefy, S., and Singer, P. (1990) Performance and Analysis of Tracer Tests to Determine Compliance of a Disinfection Scheme with the SWTR. *J. Am. Water Works Assoc.*, Dec., 88.

Trussell, R., and Chao, J. (1977) Rational Design of Chlorine Contact Facilities. *J. Water Pollut. Control Fed.,* **49,** 659.

U.S. Environmental Protection Agency (1986) *Municipal Wastewater Disinfection Manual.* EPA-625/1-86-021, Washington, D.C.

VandeVenter, L., *et al.* (1992) Chlorine Contact Time Tracer Studies Using Fluoride. *J. NEWWA,* Dec., 235.

Watson, H.E. (1908) A Note on the Variation of the Rate of Disinfection With Change in the Concentration of the Disinfectant. *J. Hyg. (G.B.),* **8,** 1908.

White, G.C. (1992) *The Handbook of Chlorination and Alternative Disinfectants.* 3rd Ed., Van Nostrand Reinhold, New York, N.Y.

SUGGESTED READINGS

Butt, J. (1980) *Reaction Kinetics and Reactor Design.* Prentice Hall, Englewood Cliffs, N.J.

Hart, F., and Gupta, S. (1978) Hydraulic Analysis of Mode Treatment Units. *J. Environ. Eng.,* **104,** 785.

Hill, C. (1977) *An Introduction to Chemical Engineering Kinetics and Reactor Design.* John Wiley & Sons, New York, N.Y.

Lev, O., and Regli, S. (1992) Evaluation of Ozone Disinfection Systems: Characteristic Time T. *J. Environ. Eng.,* **118,** 268.

Nauman, E. (1987) *Chemical Reactor Design.* John Wiley & Sons, New York, N.Y.

Qualls, R., *et al.* (1989) Evaluation of the Efficiency of Ultraviolet Disinfection Systems. *J. Water Resour., Plann. Manage. Div. Proc. Am. Soc. Civ. Eng.,* **23,** 317.

Smith, J. (1981) *Chemical Engineering Kinetics.* McGraw-Hill, Inc., New York, N.Y.

Chapter 5
Chlorination/
Dechlorination

102 Chemistry of Chlorine and Sulfur
102   Elemental Chlorine
102   Physical Properties
104   Chemical Properties
104   Toxicity
104   Hypochlorites
105   Physical Properties
105   Chemical Properties
105   Sulfur Dioxide
105   Physical Properties
106   Chemical Properties
106   Toxicity
106   Sulfite Salts
106 Chemistry and Reactions
106   Chlorine
107   Chlorine Dioxide
109   Inorganic Reactions
109   Chloramines
110   Breakpoint
111   Organic Reactions
113 Disinfection
113   Chlorine Toxicity and Effects on
          Higher Organisms
113   Aftergrowths
114   Free Versus Combined Chlorine
          Residual
116   Reduction of Chlorine Residuals
116   Dechlorination Reactions
117 Safety and Health
117   Chlorine Gas

118   Hypochlorites
118   Sulfur Dioxide
119 Shipment and Handling Safety
119   Cylinders
121   Containers
121   Facility Design
124   Ton Containers
124   Vaporizer Facilities
125   Container Hookup
128 Analytical Determination of
          Chlorine Residuals
128   Methods
128   Iodometric
128   Starch-Iodide
129   Amperometric
129   N,N-Diethyl-p-Phenylene-
             Diamine
129   Titration
129   Colorimetry
129   Leuco Crystal Violet
130   Free Available Chlorine Test
             Syringaldazin
130   Note on Orthotolidine
130   Selection of Method
131 Process Design Requirements
131   Disinfection Factors
131   Mixing
132   Closed Conduits
133   Hydraulic Devices
133   Contacting

---

135 Design and Selection of Equipment
135   Chlorinators/Sulfonators
136   Chemical-Feed Pumps
138   Sulfur Dioxide Feeders
138   Manifolds and Vacuum Regulator
           Location
140   Vaporizers
142   Residual Analyzers
143   Maintenance
144 Feed Control Strategies

145   Manual Control
145   Semiautomatic Control
146   Flow-Proportional Control
147   Residual Control
149   Compound-Loop Control
150   Cascade Control
150   Manual Injection Points
153   Dechlorination Control
155 References

CHEMISTRY OF CHLORINE AND SULFUR

Chlorine is the most widely used chemical for the disinfection of wastewater. It can be applied either as gaseous chlorine, sometimes referred to as elemental chlorine (Cl_2), or as a hypochlorite compound. Sulfur dioxide (SO_2) and other sulfur compounds, which react with excess chlorine, are used in the dechlorination of wastewater.

ELEMENTAL CHLORINE. Chlorine is an element that occurs naturally in only a combined state. At standard conditions, chlorine, a member of the halogen family, is a greenish-yellow gas. When cooled and compressed to –34.5°C (–30.1°F) and 100 kPa (one atmosphere), respectively, it condenses to a clear, amber-colored liquid. Commercial chlorine is classified as a nonflammable, toxic, compressed gas and shipped in steel containers that are designed, constructed, and handled in accordance with strict government regulations.

Physical Properties. In the gaseous state, chlorine is 2.5 times as heavy as air. In liquid form, chlorine is about 1.5 times as heavy as water. Liquid chlorine vaporizes rapidly. One volume of liquid yields approximately 450 volumes of gas. Thus 1.0 kg (2.2 lb) of liquid vaporizes to approximately 0.31 m^3 (11 cu ft) of gas.

Chlorine is only slightly soluble in water (Figure 5.1), with maximum solubility at 100 kPa (1 atm) of approximately 10 000 mg/L (1%) at 9.6°C (49.3°F). Lower temperatures will cause the formation of chlorine hydrate, a molecule of one chlorine atom and 8 water atoms, loosely bonded, that has an icelike appearance. The chlorine hydrate is easily recognized when a 100% chlorine atmosphere is present over water at low water temperatures. The appearance of chlorine ice is more a laboratory phenomenon than a common occurrence, although it once occurred regularly with the use of chlorine gas feeders referred to as bell jars. These units are seldom seen in operation today.

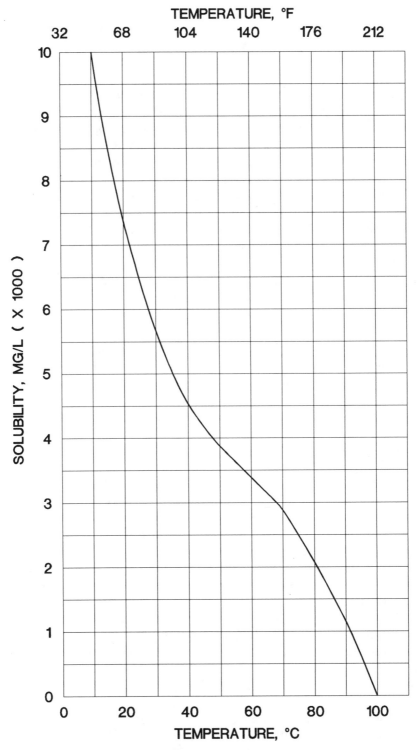

Figure 5.1   Solubility of chlorine in water (100 kPa [1 atm]) (Chlorine Institute, 1986)

The solubility of chlorine, like all gases, decreases with rising temperature (Chlorine Institute, 1986). Practical solubility is approximately 50% of theoretical.

Chemical Properties. Chlorine is highly reactive, and under specific conditions, chlorine reacts with many compounds and elements, sometimes rapidly. Because of its affinity for hydrogen, chlorine removes hydrogen from some compounds, as in the reaction with hydrogen sulfide to form elemental sulfur (S) or the sulfate ion (SO_4^{-2}), depending on the chlorine-to-sulfur ratio and reaction conditions. Chloramines are formed with ammonia or other nitrogen-containing compounds. Chlorine reacts with certain organics, compressed gases, water, and flammable materials, sometimes causing an explosion and the evolution of heat. Thus, chlorine should not be handled or stored in proximity to these materials in either the wastewater treatment plant (WWTP) or laboratory.

Most common metals are not affected at normal temperatures by dry chlorine, gas, or liquid (dry chlorine contains less than 150 ppm of water). However, chlorine reacts with titanium and ignites carbon steel at temperatures greater than 232°C (450°F).

When chlorine dissolves in water, it forms a corrosive solution containing a mixture of hydrochloric acid (HCl) and hypochlorous acid (HOCl). The corrosivity of chlorine-water solutions can create problems in handling. Most chlorine solutions are corrosive to all common metals, with the exception of gold, silver, platinum, and certain specialized alloys. Hard rubber, unplasticized polyvinyl chloride (PVC), lined metal pipe, and certain other plastics are also resistant to the corrosivity of chlorine-water solutions.

Toxicity. Chlorine gas is primarily a respiratory irritant and is classified as a toxic gas. A concentration in air greater than approximately 1.0 ppm (by volume) can be detected by most people because of its characteristic odor. Chlorine causes varying degrees of irritation of the skin, mucous membranes, and the respiratory system.

Liquid chlorine will cause skin and eye burns on contact. As noted earlier, liquid chlorine vaporizes rapidly when unconfined and produces the same effects as the gas.

Chlorine gas produces no known cumulative effects, and complete recovery can occur following mild, short-term exposures. The current Occupational Safety and Health Act permissible exposure level is 0.5 ppm, while the short-term exposure level is 1.0 ppm, and the immediately dangerous to life or health level is 30 ppm. Higher concentrations can be fatal (U.S. Code, 1993).

HYPOCHLORITES. Hypochlorites are salts of hypochlorous acid. Sodium hypochlorite (NaOCl) is the only liquid hypochlorite form in current use. There are several grades available. Calcium hypochlorite [Ca(OCl)$_2$] is the predominant dry form.

Physical Properties. Sodium hypochlorite, often referred to as liquid bleach or Javelle water, is commercially available only in liquid form, typically in concentrations between 5 and 15% available chlorine. These solutions are clear but tinted light yellow, are strongly alkaline and corrosive, and have a strong chlorinous odor.

Calcium hypochlorite, sometimes referred to as powder bleach, is a dry material typically consisting of 65% available chlorine. Often called high-test hypochlorite (HTH), 1 kg (2.2 lb) is equivalent to 0.65 kg (1.43 lb) of elemental chlorine. This solid is a white, hygroscopic material that emits a strong, chlorinous odor.

Chemical Properties. Hypochlorites are strong oxidants. All sodium hypochlorite solutions are unstable. Heat, light, storage time, and impurities such as iron accelerate product degradation. Hypochlorites are destructive to wood, corrosive to most common metals, and will adversely affect the skin, eyes, and other body tissues with which they come in contact.

The dry form (calcium hypochlorite) is unstable under normal atmospheric conditions. Reactions may occur spontaneously with numerous chemicals including turpentine, oils, water, and paper. Calcium hypochlorite, therefore, should be stored in dry locations and used only with equipment free of organics. Serious fire and explosion hazards exist when using this material.

SULFUR DIOXIDE. Sulfur dioxide is classified as a nonflammable, corrosive, liquefied gas and is shipped commercially in steel containers designed, constructed, and handled in accordance with strict government regulations.

Physical Properties. In the gaseous state, sulfur dioxide is colorless with a suffocating, pungent odor and is approximately 2.25 times as heavy as air. Liquid sulfur dioxide is approximately 1.5 times as heavy as water. Commercially, sulfur dioxide is supplied as a pressurized, colorless, liquefied gas. The gas solubility in water (approximately 20 times greater than that of chlorine) is approximately 18.6% at 0°C (32°F). In solution, sulfur dioxide hydrolyzes to form a weak solution of sulfurous acid (H_2SO_3). Sulfurous acid dissociates according to Equations 5.1 and 5.2:

$$H_2SO_3 \leftrightarrow H^+ + HSO_3^-, \qquad pK_a = 1.76 \qquad (5.1)$$

$$HSO_3^- \leftrightarrow H^+ + SO_3^{-2}, \qquad pK_a = 7.19 \qquad (5.2)$$

Where the dissociation constants are at 25°C (77°F).

At a pH greater than 8.5, 95% of the sulfur dioxide gas dissolved in water exists as the sulfite ion (SO_3^{-2}). The solubility of sulfur dioxide in water decreases at elevated temperatures.

Chlorination/Dechlorination   *105*

Chemical Properties. Dry sulfur dioxide, liquid or gas, is not corrosive to steel and most other common metals. However, galvanized metals should not be used to handle sulfur dioxide, and in the presence of sufficient moisture, sulfur dioxide is corrosive to most common metals. It is neither flammable nor explosive in either the gaseous or liquid state. Because sulfur dioxide neither burns nor supports combustion, there is no danger of fire or explosion.

Toxicity. Sulfur dioxide is an extremely irritating gas. The gas may cause varying degrees of irritation to the mucous membranes of the eyes, nose, throat, and lungs because of sulfurous acid formation. Higher concentrations of sulfur dioxide over a prolonged period produce a suffocating effect. Contact with the liquid results in freezing of the skin because the liquid absorbs its latent heat of vaporization from the skin. Worker exposures, on an 8-hour, time-weighted average, are currently limited by the Occupational Safety and Health Administration (OSHA) to 5 mg/L by volume in air, or approximately 13 mg/m^3. Concentrations of 500 mg/L are acutely irritating to the upper respiratory system and cause a sense of suffocation after several inhalations (Boliden Intertrade, 1979, and Compressed Gas Association, 1988).

SULFITE SALTS. Sodium sulfite (Na_2SO_3), sodium bisulfite ($NaHSO_3$), and sodium metabisulfite ($Na_2S_2O_5$) are also used in dechlorination. On dissolution in water, these salts produce the same active ion, sulfite (SO_3^-). The salts are available as dry chemicals and should be handled with the same care given to sulfur dioxide. All three compounds are generally more expensive than sulfur dioxide per pound of active reducing agent, but of the three, sodium metabisulfite is less costly and more stable than sodium sulfite and sodium bisulfite.

*C*HEMISTRY AND REACTIONS

CHLORINE. When chlorine is added to water or wastewater, hydrolysis occurs, and a mixture of hypochlorous acid and hydrochloric acid is formed. The reaction (Equation 5.3) is pH and temperature dependent, completed within milliseconds, and reversible. The hypochlorous acid formed is a weak acid and dissociates or ionizes to form an equilibrium solution of hypochlorous acid and hypochlorite ion (OCl^-) (Equation 5.4). The equilibrium approaches 100% dissociation ($H^+ + OCl^-$) when the pH exceeds 8.5 and approaches 100% hypochlorous acid when the pH is less than 6.0.

$$Cl_2 \ (g) + H_2O \ \leftrightarrow \ HOCl + H^+ + Cl^-, \qquad (5.3)$$
$$pK_a = 7.537914 \ @ \ 25°C \ (77°F)$$

$$HOCl \ \leftrightarrow \ H^+ + OCl^-, \quad pK_a = 7.537 \ @ \ 25°C \ (77°F) \qquad (5.4)$$

Chlorine exists predominately as hypochlorous acid at pH levels between 2 and 6. At a pH of less than 2, hypochlorous acid reverts to chlorine; at a pH of greater than 7.8, hypochlorite ions are the predominate form. This distribution is illustrated on the graph of pH shown in Figure 5.2.

Hypochlorite solutions of sodium hypochlorite and calcium hypochlorite also dissociate in water to form hypochlorite ions (Equations 5.5 and 5.6). The formation of both hydrogen and hypochlorite ions is pH dependent. Just as exhibited in the hydrolysis of chlorine gas, an equilibrium is established. The hypochlorite solutions formed are at an elevated pH. The equilibrium curve (Figure 5.2) also applies to solutions of sodium and calcium hypochlorite. Chlorine tends to decrease the pH, while the hypochlorites tend to raise the pH. Chlorine also reduces the alkalinity by as much as 2.8 parts of chlorine per part of calcium carbonate ($CaCO_3$). An alkali is typically added to calcium and sodium hypochlorite solutions to enhance stability and slow decomposition. Thus, hypochlorite solutions can have pHs greater than 10. The treated wastewater is not typically affected by the pH of the chlorine solution added regardless of the source of the chlorine (chlorine gas or hypochlorites).

$$NaOCl \leftrightarrow Na^+ + OCl^- \qquad (5.5)$$

$$Ca(OCl)_2 \leftrightarrow Ca^{+2} + 2OCl^- \qquad (5.6)$$

The chlorine present as both hypochlorous acid and hypochlorite ions is defined as free available chlorine (FAC). Aqueous chlorine may also be referred to as FAC, but its usefulness is not practical because of the low pH conditions involved.

CHLORINE DIOXIDE. Chlorine dioxide (ClO_2), a chlorine derivative, is a proven, well-recognized disinfectant and has excellent bactericidal and virucidal properties. Chlorine dioxide has been used in wastewater treatment for specialized wastes such as those containing phenols or sulfides. The use of chlorine dioxide in drinking water is becoming more widespread because of its low CT values and its ability to oxidize the organics that can lead to carcinogenic compound formations such as trihalomethanes (THMs). Chlorine dioxide has also been used extensively in bleaching applications at pulp and paper mills to aid in the prevention of dioxin formation. Regardless of these successful installations and its easy availability, its use in wastewater treatment for disinfection is limited. Generally, the economics involved in generating chlorine dioxide and the benefits obtained compared to chlorine influence the decision not to use it. Therefore, chlorine dioxide will not be discussed in detail in this manual. The reader is directed to manuals and texts dealing specifically with this topic (Vulcan Chemicals, 1993).

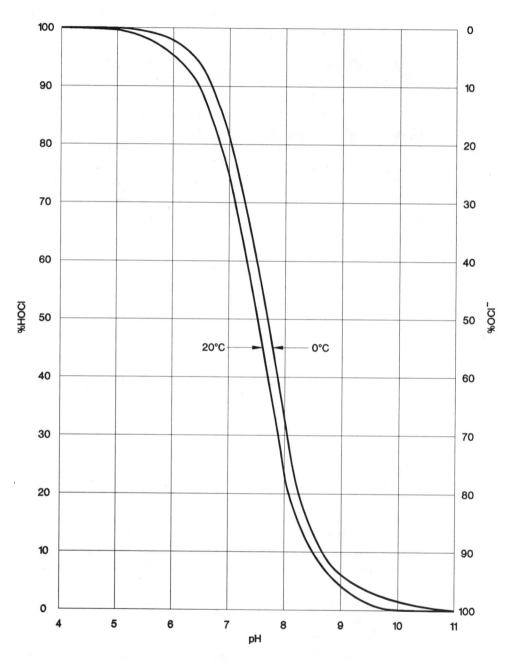

Figure 5.2 Chlorine, hypochlorous acid: hypochlorite distribution versus pH (reprinted from *Water Chlorination Principles and Practices*, M20, by permission; copyright © 1973, American Water Works Association)

INORGANIC REACTIONS. Inorganic compounds such as hydrogen sulfide (H_2S), sulfur dioxide, and sulfite salts (SO_3^{-2}) can react with oxygen present in the waste stream. These compounds can be associated with septic conditions of the wastewater. Nitrite (NO_2^-), ferrous iron (Fe^{+2}), and manganous manganese (Mn^{+2}) can also be found in waste streams. These compounds have a chlorine demand and are typically defined as reducing agents because they react rapidly with the free chlorine, thereby reducing the effective chlorine amounts. Inorganic reducing agents typically react easily and rapidly with the oxidizing agents hypochlorous acid and/or hypochlorite ion.

When chlorine is added to water containing hydrogen sulfide or sulfite compounds, these compounds are oxidized to elemental sulfur or sulfate, depending on the conditions of the reaction. Nitrite is oxidized to nitrate (NO_3^-), ferrous to ferric (Fe^{+3}), and manganous to manganic (Mn^{+4}) ions. The chlorine product in the reaction is the chloride ion, a stable, nonbactericidal material.

Chloramines. Chloramines are products of the reactions between chlorine and ammonia-nitrogen found in wastewater. Ammonia-nitrogen may be present in appreciable amounts, typically 10 to 40 mg/L, in wastewater as either dissolved ammonia gas (NH_3), the ammonium ion (NH_4^+), or organic nitrogen compounds. The ammonia-ammonium equilibrium (Equation 5.7) is pH dependent:

$$NH_3(g) + H_2O \leftrightarrow NH_4^+ + OH^-,$$
$$pK_a = 9.24 \ @ \ 25°C \ (77°F) \tag{5.7}$$

The most significant source of ammonia is the urea (Equation 5.8) present in urine and the hydrolysis products of proteins. A small amount of ammonia-nitrogen may come from industrial discharges, agricultural runoff, or fertilizer leachate. Equation 5.8 demonstrates the formation of ammonium, which then equilibrates with ammonia according to Equation 5.7.

$$CH_4N_2O + 3H_2O \rightarrow 2NH_4^+ + HCO_3^- + OH^- \tag{5.8}$$

Three chloramines are formed—monochloramine (NH_2Cl), dichloramine ($NHCl_2$), and trichloramine or nitrogen trichloride (NCl_3)—in a stepwise process beginning with the reaction of hypochlorous acid and ammonia in dilute aqueous solutions (Equations 5.9, 5.10, and 5.11). Chlorine and ammonia can also react (Equation 5.12), although this reaction should not be considered a representation of the reaction mechanism.

$$NH_3 + HOCl \leftrightarrow NH_2Cl + H_2O \tag{5.9}$$

$$NH_2Cl + HOCl \leftrightarrow NHCl_2 + H_2O \qquad (5.10)$$

$$NHCl_2 + HOCl \leftrightarrow NCl_3 + H_2O \qquad (5.11)$$

$$2NH_3 + 3HOCl \leftrightarrow N_2\hat{\ }(g) + 3HCl + 3H_2O \qquad (5.12)$$

How far these stepwise reactions proceed, and how much of each compound is formed, depends on the pH, temperature, time of contact, and reactant (ammonium ion and hypochlorous acid) concentrations. All four reactions (Equations 5.9 to 5.12) may take place simultaneously and compete with one another. In general, low pH levels and high chlorine-to-ammonia ratios favor dichloramine ($NHCl_2$) formation. At a pH of greater than approximately 8.5, monochloramine (NH_2Cl) exists almost exclusively. At a pH between 8.5 and 5.5, monochloramine and dichloramine exist simultaneously; between a pH of 5.5 and 4.5, dichloramine exists almost exclusively. At a pH of less than 4.4, nitrogen trichloride will be produced (Jafvert and Valentine, 1992).

Organic chloramines are formed when chlorine reacts with amino acids, proteinaceous material, and other organic nitrogen forms. Organic chloramines are not effective disinfectants, and the presence of organic nitrogen can exert considerable chlorine demand in addition to interfering with differentiation between free and combined forms. Chlorine existing in chemical combination with ammonia or organic nitrogen chloramines is termed *combined available chlorine* (CAC).

Breakpoint. Breakpoint chlorination is the process of using chlorine's oxidative capacity to oxidize ammonia to nitrogen (Equations 5.9 to 5.12). Although breakpoint chlorination is a practice used in wastewater treatment to obtain a free chlorine residual in the presence of ammonia-nitrogen (NH_3–N), it is not used in conventional wastewater treatment because of the large quantities of chlorine required (approximately a 10:1 mass ratio of chlorine to ammonia-nitrogen). An exception to this occurs when a WWTP knowingly or unknowingly nitrifies most of its ammonia-nitrogen, leaving little or no ammonia present to form chloramines. In such a case, the addition of chlorine can produce the breakpoint reaction. In breakpoint chlorination, ammonia is progressively oxidized until a point beyond which the combined chlorines or chloramines and ammonia react with chlorine to produce nitrogen gas. In this case, ammonia is no longer present as ammonia gas or ammonium ion, and only a minimum is present in combined chlorine forms.

The reaction (Equations 5.9 and 5.10) proceeds at a rate that is highly pH sensitive. At pH 7 to 8, and when the mass ratio of chlorine to ammonia-nitrogen is 5:1 or less, all free chlorine is converted to monochloramine. At a lower pH, dichloramine is formed. When the applied chlorine-to-ammonia-nitrogen ratio exceeds 5:1, the total chlorine residual decreases and the breakpoint is approached, with N_2 as a product. When the breakpoint is reached (a chlorine-to-ammonia mass ratio from 8:1 to 10:1), the ammonia-nitrogen is oxidized, and free chlorine begins to appear in small amounts in the resultant residual. The reaction also slows appreciably with lower temperatures. The presence of organic nitrogen influences the shape of the breakpoint curve and appears as an organic chloramine after the breakpoint. Chlorine may have relatively little influence on organic nitrogen as compared to ammonia-nitrogen (Figures 5.3 and 5.4). The presence of organic nitrogen influences the shape of the breakpoint curve and appears as an organic chloramine after the breakpoint. Chlorine may have relatively little influence on organic nitrogen as compared to ammonia-nitrogen.

ORGANIC REACTIONS. Chlorinated organic compounds, or chloro-organics, are formed by the action of chlorine on the total organic carbon contained in wastewater, just as organic chloramines are products of the chlorination of the total organic nitrogen fraction. Organic compounds are those that contain carbon atoms, either singularly, in chains, or in ring structures. Examples of organic compounds are methane (natural gas), coal, petroleum, alcohol, and acetic acid (vinegar). Organic nitrogen compounds contain nitrogen atoms in addition to carbon atoms. Examples of organic nitrogen compounds include urea, amino acids, and proteins.

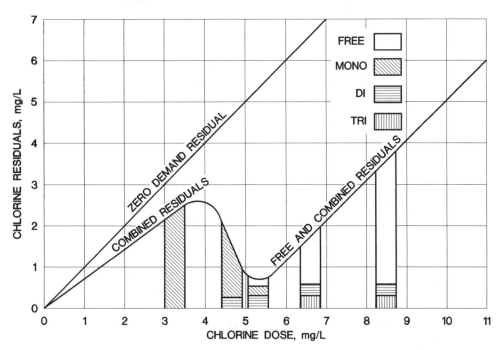

Figure 5.3   Chlorine residual with ammonia nitrogen (from White, G.C. [1992] *The Handbook of Chlorination and Alternative Disinfectants*. 3rd Ed., Van Nostrand Reinhold, New York, N.Y., with permission)

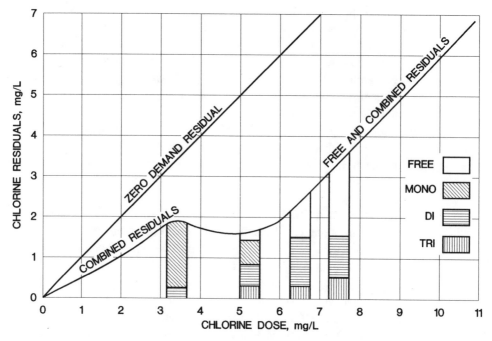

Figure 5.4 Chlorine residual with ammonia nitrogen and organic nitrogen (from White, G.C., [1992] *The Handbook of Chlorination and Alternative Disinfectants*. 3rd Ed., Van Nostrand Reinhold, New York, N.Y., with permission)

Trihalomethanes are the products of complex chemical reactions of chlorine with a group of organic acids known as *humic* acids. The results are single carbon molecules containing halogen atoms present in varying combinations. Also known as *haloforms*, THMs can be represented diagrammatically as

$$HCX_3 \qquad (5.13)$$

Where X is either a Cl^- or Br^- atom (for example, $HCCl_3$ [chloroform]).

The THM group of compounds also includes chlorodibromo methane, bromodichloro methane, and bromoform. There is a concern with the formation of THMs because of their environmental impact and effect on human health. Chloroform (Equation 5.13), a well-known THM, is a documented animal carcinogen, and all haloforms are believed to act in a similar manner. The U.S. Environmental Protection Agency (U.S. EPA) has issued regulations for the drinking water industry (Safe Drinking Water Act) targeted at minimizing public exposure to this class of compounds. Current regulations call for a maximum trihalomethane concentration of 0.1 mg/L in drinking water. Recently developed regulations will set THM standards for drinking water at 0.08 mg/L by 1997.

Precursors, organic molecules from which THMs are formed, are typically from the organic acid groups and are known to be contained in wastewater. Although discharges from WWTPs may contain more precursors than the receiving water, the quantity of THMs present in wastewater effluent may actually be small because of the ammonia reaction selectivity and the slow reaction to form THMs in the presence of either chlorine form, FAC or CAC.

DISINFECTION

CHLORINE TOXICITY AND EFFECTS ON HIGHER ORGANISMS.
Although the need for disinfection is discussed in Chapter 3, there are some aspects of chlorine disinfection that should be mentioned here. Chlorine is an extremely reactive element, rapidly undergoing chemical reactions with inorganic and organic substrates. When the organic substrate is part of a living organism, the reaction can have a toxic effect on the organism. This toxicity may affect the organism's ability to reproduce or metabolize, cause genetic dysfunctions (mutations), or ultimately kill the organism. The active agents are hypochlorous acid, hypochlorite ion, monochloramine, and dichloramine. Hypochlorous acid reacts with organic nitrogen compounds and may alter the chemical structure of the organism's organic materials and change the genetic information.

Chlorine, when added to wastewater as a disinfectant, has varied effects both on downstream users of the receiving water body and on the living organisms that may present in the waste effluent or the biota of the receiving water. In the waste effluent, cysts of *Entamoeba histolytica* and *Giardia lamblia* and eggs of parasitic worms are resistant to chemical disinfectants. Because of their high resistance to disinfection, cysts and ova can be removed from wastewater more effectively by methods other than chlorine disinfection. Coagulation/sedimentation followed by filtration is the primary method of removing these organisms from potable water. Chlorination has limitations, and wastewater treatment operators should not try to accomplish by disinfection what can be accomplished more effectively and economically by other methods. Typical reductions in virus concentrations from three treatment levels are given in Table 5.1.

AFTERGROWTHS. In the receiving stream, aftergrowths of some organisms occur after discharge of chlorinated effluents. These aftergrowths have two particularly undesirable effects: they often lead to serious oxygen deficiencies in receiving waters and a substantial aftergrowth of coliforms. The extent of these effects is primarily governed by the amount of biologically oxidizable material in the effluent. The aftergrowths observed in waters receiving chlorinated effluents are presumed to be a result of the destruction of large numbers of protozoa by chlorination. This permits subsequent multiplication of the surviving bacteria unhampered by predatory protozoa such as

Table 5.1   Expected virus concentration in effluent

Treatment	Removal expected, %	Virus concentration in effluent, no./L[a]
Primary sedimentation (without chemicals)	0	7 000
Secondary treatment		
Trickling filters	50	3 500
Activated sludge	90	700
Physical/chemical treatment		
Precipitation of phosphate and suspended solids	90	
Activated carbon adsorption	10	630

[a]   Assumes a virus concentration of 7 000/L in the raw wastewater.

the ciliates and flagellates. The aftergrowths develop rapidly after chlorination, presumably because of the abundant food supply available for the small number of surviving organisms combined with the absence of predatory protozoa. When effluents are chlorinated, the greater the initial reduction in bacterial population, the longer the lag time will be before the multiplication of surviving organisms becomes apparent.

The undesirable effects of chlorination on the receiving water biota have been documented. Combined chlorines may be present in the effluent, and because this form of chlorine is less active and more persistent, its presence can cause fish kills and other disruptions in the ecosystem. Dechlorination, most often with sulfur dioxide, has been demonstrated to be effective in reducing this problem. Most states now require dechlorination, and U.S. EPA requirements call for dechlorination to levels of total chlorine less than 0.05 mg/L in effluent discharged to receiving water bodies. However, downstream drinking-water plants consider the chlorination of effluent a helpful additional barrier to the transfer of disease from upstream discharges.

FREE VERSUS COMBINED CHLORINE RESIDUAL. Toxicity is related to the individual sensitivities of the target organisms. A residual's toxicity is also related to the degree of chemical reactivity of the compound. Free chlorine residual, hypochlorous acid, and hypochlorite ion are more reactive compounds than are the combined residuals, mono- and dichloramine. Figure 5.5 relates residual concentration and contact time to 99% destruction of *Escherichia coli,* the typical coliform. The data in Figure 5.5 were derived from pure-water studies. Although wastewater would require significantly higher chlorine doses, the relationship between the relative efficiencies of the various residuals remains.

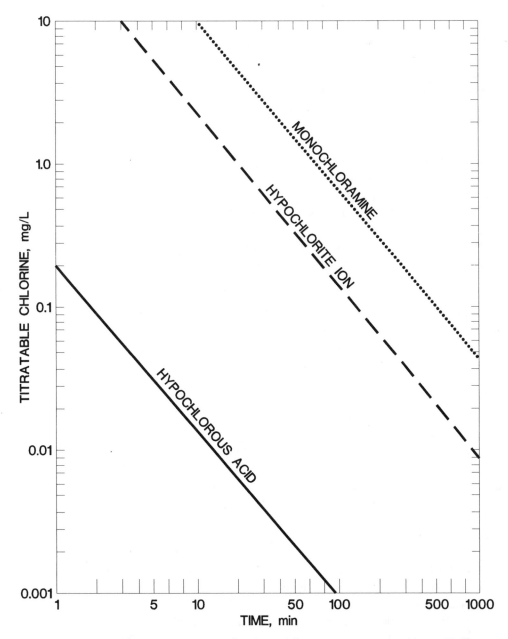

Figure 5.5 *Escherichia coli* kill times versus residual concentration (from Clarke, N.A., *et al.* [1964] *Human Enteric Viruses in Water: Source, Survival and Removability, Advances in Water Pollution Research.* Vol. 2, Pergamon Press, London, U.K., 523)

Free chlorine residuals, hypochlorous acid and hypochlorite ion, exist only momentarily when added to wastewater containing ammonia-nitrogen. Because ammonia is a major constituent of most effluents and chlorine is able to form chloramines (Equations 5.9, 5.10, and 5.11), it follows that combined residuals, which generally are predominately monochloramine, will be responsible for most of the germicidal activity of chlorine present in chlorinated effluents.

Free chlorination does not necessarily lead to improved disinfection. Presumably, the reactive free residual is dissipated in organic reactions and no longer available for disinfection. The combined residuals do not undergo such side reactions and remain available for disinfection.

Free residuals are more reactive than combined residuals, although they may be consumed in organic reactions that do not contribute to disinfection. Free chlorine can produce undesirable byproducts that may be harmful to humans or to the biota of receiving streams, while combined residuals can kill biota in the receiving stream.

REDUCTION OF CHLORINE RESIDUALS. Much attention has been focused on the toxic effects of chlorinated effluents. Both free chlorine and chloramine residuals are toxic to fish and other aquatic organisms even at concentrations less than 0.02 mg/L. Although fish are repelled by low levels of chlorine and frequently escape harm, other aquatic organisms in the food chain may be killed by chlorine discharges. Dechlorination, the removal of the remaining chlorine, is required in most states. Stream standards have been established in most parts of the country that limit chlorine as total residual. In Maryland, the level has been established at 0.02 mg/L for trout streams and for waters discharging into the Chesapeake Bay. Many states are considering the use of ultraviolet light for disinfection, which would supply a chlorine-free residual. The reader should contact the appropriate state regulatory agency to determine the current acceptance in specific states. Ultraviolet treatment is covered in Chapter 7.

DECHLORINATION REACTIONS. Free and combined chlorine residuals can be effectively reduced by sulfur dioxide and sulfite salts. The sulfite ion reacts rapidly with free and combined chlorine.

The sulfite ion is the active agent when sulfur dioxide or sulfite salts are dissolved in water. Their dechlorination reactions are identical. Sulfite reacts instantaneously with free and combined chlorine (Equations 5.14 and 5.15):

$$SO_3^- + HOCl \rightarrow SO_4^{-2} + Cl^- + H^+ \qquad (5.14)$$

$$SO_3^{-2} + NH_2Cl + H_2O \rightarrow SO_4^{-2} + Cl^- + NH_4^+ \qquad (5.15)$$

The reactions yield small amounts of acidity, which is neutralized by the alkalinity of the wastewaters (2.8 mg of alkalinity as calcium carbonate is consumed per milligram of chlorine reduced). From the above equations, the amount of sulfur dioxide required per part of chlorine is 0.9, but the actual practice calls for the use of a 1:1 ratio.

Sulfur dioxide and sulfite salts may also react with dissolved oxygen to produce sulfates. Although excess sulfite could lead to a reduction in dissolved oxygen in the effluent and the receiving stream, the amount of sulfite in a properly controlled system is insignificant. Approximately four parts of sulfite are required per part of oxygen removed.

Granular and powdered carbon may be used to dechlorinate free, and some combined, chlorine residuals. Their use is generally limited to specific sites or effluents with special discharge limitations. In addition, carbon provides filtration that removes other undesirable materials.

Carbon requirements for dechlorination are typically determined by on-site pilot testing. Parameters of significance include mean particle diameter of the carbon (pressure drop within a contactor) and influent quality (pH, organics, and colloids). Doses in the range of 30 to 40 mg/L have been reported.

Storage of the final chlorinated effluent before final discharge can effectively reduce the chlorine concentration. While this practice tends to be land intensive, secondary benefits, such as the small amount of additional treatment required, may be realized.

SAFETY AND HEALTH

CHLORINE GAS. Chlorine gas has a detectable odor at low levels of concentration and has a greenish yellow color at higher levels of concentration. At volumes of less than 0.1 ppm in air, chlorine gas is undetectable except by instruments. The maximum contaminant level established for chlorine gas by OSHA is a 1-ppm time-weighted average over 8 hours. Chlorine levels can also be detected using ammonia vapor, but the detection is only qualitative. In this method, a vapor spray from a strong (26° Baume) aqueous solution of commercial-grade ammonia directed to the vicinity of the suspected leak will, in the presence of chlorine vapor, produce a white cloud of ammonium chloride.

Harmful effects of chlorine gas exposure begin to manifest at approximately 5 ppm and higher. Between 5 and 10 ppm, however, these effects (choking, coughing, watery eyes, mild skin irritation, and lung irritation) are temporary. At higher concentrations, the effects become more long lasting and can result in death.

All of the commonly used forms of chlorine are hazardous chemicals. The precautions necessary for each chemical are different. Wastewater treatment plant management personnel should develop and practice safety procedures

---

including the use of self-contained breathing apparatus, repair kits, neutralization procedures, and evacuation plans and should involve other local agencies such as fire, police, and health and emergency medical services in this planning. Personnel handling chlorine must be adequately trained. The chemical supplier and other equipment manufacturers will typically provide personnel and materials to aid in such training. The Chlorine Institute also makes training films and information available for this purpose.

High-capacity storage and transportation facilities should be located in isolated areas and equipped with scrubbers. The designer should consult the latest edition of *The Chlorine Manual* (Chlorine Institute, 1986), Compressed Gas Association data, the appropriate Department of Transportation and OSHA regulations, and any local fire codes such as the Uniform Fire Code (*Uniform Fire Code,* 1994), Standard Fire Prevention Code (Southern Building, 1991), and The Building Officials and Code Administration International's *National Fire Code* (BOCA, 1993). Department of Transportation regulations govern the use of tank cars, trucks, and ton (910-kg) containers. Local codes and regulations do not govern these areas.

HYPOCHLORITES. Eye protection and access to an emergency eyewash and showers are recommended for operators handling sodium hypochlorite. As with any form of hypochlorite, the undiluted chemical can cause severe burns on the skin and clothing; it is recommended that operators working with either of the hypochlorites wear protective clothing.

Operators who use calcium hypochlorite should always wear eye protectors and dust masks when transporting the powder or mixing it with water. In addition, they should always have immediate access to an eyewash and shower facility when working with the chemical. All areas exposed to hypochlorite should be washed thoroughly. If pressed HTH discs are used, rubber gloves are recommended to provide hand protection.

SULFUR DIOXIDE. Sulfur dioxide, a gas classified as a corrosive chemical, is an irritant and much more soluble than chlorine in water. Sodium sulfite and bisulfite, typically provided in solution form, are reducing agents like sulfur dioxide. Both solutions are corrosive and must be stored in containers similar to those used for hypochlorite solutions.

Detailed information on safe handling procedures for all chemicals are available from the chemical supplier. Training of all personnel in the handling and use of sulfur dioxide is available from the Compressed Gas Association and individual manufacturers (CGA, 1988).

SHIPMENT AND HANDLING SAFETY

Operators of WWTPs in which hazardous chemicals are used must be thoroughly familiar with U.S. Department of Transportation regulations regarding the transportation of these chemicals. Calcium hypochlorite is classified as a corrosive and rapid oxidant. Sodium hypochlorite is a corrosive agent. Chlorine and sulfur dioxide are nonflammable, corrosive, toxic, liquefied gases under pressure. The various sulfite solutions are classified as corrosive.

Shipment manifests are the responsibility of the transporter and must be kept every time any quantity of chemical is transported. If the wastewater treatment facility is transporting chemicals between WWTPs, it is responsible for keeping such manifests.

CYLINDERS. The following precautions for chlorine or sulfur dioxide containers should be observed during transport, delivery, and use:

- Vehicles carrying more than 450 kg (1 000 lb) of chemicals must carry the standard diamond-shaped U.S. Department of Transportation warning signs, regardless of whether they are commercial carriers or agency vehicles.
- Gas cylinders must be secured during shipment by means of chains and transported in an upright position. Horizontal containers (ton [910 kg]) must be secured by block and chain or straps.
- The truck motor should be shut down during unloading.
- All hoists must be rated for the full load, including the weight of the empty containers and lifting tackle. The cables or chains must not be frayed or damaged.
- A 70-kg (150-lb) cylinder should never be lifted by the valve or valve bonnet. All cylinder and ton container valves must be protected during shipment and handling.
- Containers must never be moved without the safety cover in place.
- Containers must never be dropped or struck.
- Storage areas must be posted properly with signs in accordance with local, state, and federal laws and regulations.
- Chlorine and sulfur dioxide should be stored in a divided or separate room.
- The temperature of the cylinders must never exceed or approach 70°C (158°F). This is the temperature at which the fusible plugs are designed to melt, which prevents hydrostatic rupture.
- Water may be sprayed on containers to keep them cool in the event of a fire. Water should never be sprayed on leaking containers because the presence of water will increase the corrosion of the metal and escalate the leak.

- All containers must be stored in a well-ventilated area, away from external heat sources and direct sunlight.
- Self-contained breathing apparatus of the pressure-demand type and a Chlorine Institute cylinder repair kit of the appropriate type ("A" for 70-kg [150-lb] cylinders, "B" for ton containers, and "C" for tank trucks, cars, and barges) must be available for immediate use in case of a leak. Kits for sulfur dioxide are the same as those for chlorine with the exception that gaskets provided in the kits are a different material for each gas.
- Canister-type gas masks may not be used for protection against chlorine or sulfur dioxide inhalation in oxygen-deficient areas.
- An emergency escape respirator should always be on hand when working with chlorine or sulfur dioxide cylinders.
- Use of personal foot guards during cylinder handling and hard hats during ton container handling when using hoists is strongly recommended.
- The BOCA *National Fire Code* (BOCA, 1993) and Standard Codes (Southern Building, 1991) permit the use of ton containers and Chlorine Institute emergency kits without the need for treatment systems (scrubbers).
- There are three codes that have been developed as models for use by states, counties, and municipalities. The codes are not legally binding unless they are enacted into law by the appropriate authorities. The Uniform Code, developed by the Western Fire Chief's Association based in Whittier, California, is generally used by states west of the Mississippi (*Uniform Fire Code,* 1994). The Standard Code, developed by the Southeastern Fire Chief's Association and the Southwestern Fire Chief's Association based in Birmingham, Alabama, is generally used by states south of the Mason-Dixon Line and east of the Mississippi, plus Texas and Louisiana (Southern Building, 1991). The *National Fire Code*, developed by the Building Officials and Code Administrators of America based in Chicago, Illinois, is generally used by the states north of the Mason-Dixon Line and east of the Mississippi (BOCA, 1993). These areas are approximate, and each state code should be reviewed.
- Current regulations for all three codes permit up to four 70-kg (150-lb) cylinders per control area without the use of treatment systems (scrubbers) if the control area is equipped with sprinklers and has a 1-hour fire rating. Up to four control areas are permitted per building. The applicable fire codes should be reviewed, and the applicable local regulations should be determined to provide the proper installation.

- Chlorine and sulfur dioxide gas detectors are recommended for all storage and use areas. Local alarm lights or audible devices are typically used. For unmanned facilities, connection of the alarm to an external manned location is recommended.

CONTAINERS. Chlorine and sulfur dioxide are supplied in steel pressure vessels (Figure 5.6) of 45 and 70 kg (100 and 150 lb) and in ton containers (Figure 5.7), tank trucks and tank cars, and barges. For large WWTPs, a rail siding or stationary bulk storage tank may be provided.

The 70-kg (150-lb) cylinders are moved, stored, and used in the upright position. The cylinder valve is equipped with a fusible plug in its body. This plug is designed to melt at between 70 and 73.9°C (158 and 165°F) to prevent hydrostatic rupture of the cylinder.

Ton containers are moved, stored, and used in a horizontal position. The container valve is similar to the upright standard cylinder valve except that it has no fusible plug. Three fusible plugs are located on each end of the ton container. Each valve on a ton container is connected to a tube inside the container. The valves on these containers must be aligned in the vertical position to enable gas withdrawal from the upper valve and liquid withdrawal from the lower valve (Figure 5.7). The Chlorine Institute's recommended orientation of the upper valve is to have the valve outlet facing to the right. Because this recommendation is not mandatory, some chlorine packagers do not use this valve orientation. Some have the upper valve outlet facing to the left. It is important that all WWTP personnel know what the supplier is providing. If direct mounting of vacuum regulators is in practice at the facility, the orientation may be critical to some types of equipment.

Tank trucks of 13 500 to 18 000 kg (15 to 20 ton) capacity are available, as are single-unit tank rail cars of 14 500, 27 000, 50 000, 77 000, and 81 500 kg (16, 30, 55, 85, and 90 ton) capacity.

FACILITY DESIGN. All handling facilities must be designed with adequate space for loading and unloading cylinders or ton containers. Vehicles used to transport cylinders must be equipped with an upright cylinder rack, chain restraints for the cylinders, and a lift gate. If no lift gate is available on the vehicle, the facility should be constructed with a raised loading dock and interior ramp to allow the cylinders to be transported without lifting by hand.

A properly designed handling facility will have panic or escape hardware on all doors. All entrances from storage or use areas should be from outside the building, and all doors should be designed to open outward. Every room in a chlorine or sulfur dioxide facility in which a chlorine or sulfur dioxide gas container is stored or used must be negatively ventilated with one air change per minute and have a gas detector. Because these gases are heavier than air, an exhaust blower and screened floor vents are recommended. The gas detector should be interlocked with the fan, as well as with audible or visible

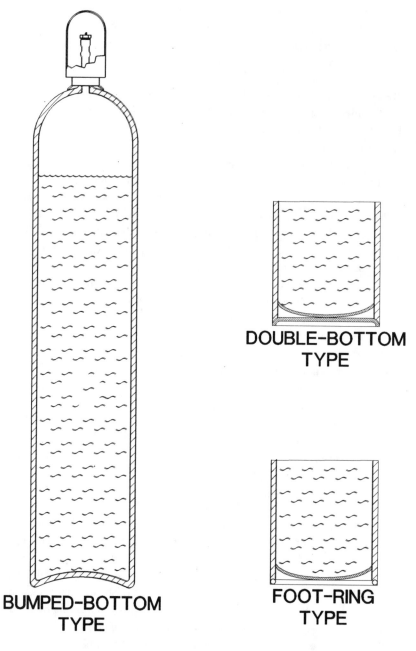

BUMPED-BOTTOM TYPE

DOUBLE-BOTTOM TYPE

FOOT-RING TYPE

Figure 5.6   Types of 70-kg (150-lb) cylinders

alarms. Proper design would also permit observation of the gas detector from
outside the storage room. Currently available gas detectors use remote sensors
that permit the indicator/transmitter to be installed outside the storage area and
as far as 330 m (1 000 ft) away.

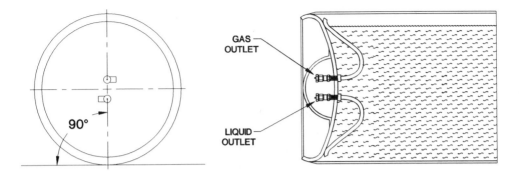

Figure 5.7   Ton container orientation for proper use

The Ten States Standards (1992) are a good reference for facility design. These standards were developed by a group of public health officials from the U.S. states roughly bordering the Great Lakes. The province of Ontario, Canada, was also an active participant. The standards are revised periodically to include current data.

The doors of the facility should have an electrical interlock that automatically turns on the lights and exhaust fan in the room before entry and when the doors are opened. A manual switch to operate the lights and fan should also be located near the doors on the outside wall. A gas-tight window should be installed in at least one of the doors of the facility to permit observation of the interior before entry. The ready availability of exterior emergency shower and eyewash stations should also be considered.

Mounted on the front wall outside the facility near the door should be the self-contained breathing apparatus. The appropriate OSHA-approved warning signs should be posted at the entrance and any other exposed side. The appropriate container repair kit should be located at a convenient external location.

The gas storage and use areas should be dedicated rooms. In these facilities, nothing should be stored and no work performed that is not related directly to handling chlorine or sulfur dioxide. The chlorine storage and chlorine feeder rooms must be separate and well lighted, with adequate maneuvering area. There should be no common drain, ventilation system, or doors between them.

A facility handling 70-kg (150-lb) cylinders should be equipped with at least one cylinder dolly. Cylinders should not be rolled across the ground on the bottom rim. Cylinder racks or suitable support must be provided for all cylinders, and the racks must be securely attached to the floor or wall of the facility.

Facilities handling chlorine and sulfur dioxide should be supplied with adequate heating systems to achieve temperature levels between 18 and 43°C (64.4 and 110°F). This is especially important for sulfur dioxide systems because maximum withdrawal of sulfur dioxide gas may be achieved only at temperatures of 18°C (64.4°F) or higher. If this temperature is not desirable, withdrawal from multiple cylinders or the use of a separate vaporizing system should be considered.

It may be desirable to have a workbench, tools, and parts that are necessary to maintain the chlorination or sulfonation equipment in or near the storage/use rooms.

In any facility, it is recommended that chlorine or sulfur dioxide equipment be located in isolated rooms separate from the cylinder storage. Remote vacuum regulators should be used to convert gas to a vacuum, preferably at the source. This will maintain any gas piping in the equipment room under vacuum, which improves safety, and will permit the use of plastic piping such as schedule 80 PVC from the regulator to the feeding equipment and ejector. To facilitate operation, a gas-tight observation window should be provided in the common wall between the storage/vacuum regulator rooms.

TON CONTAINERS. Facilities handling ton containers must meet the same requirements of adequate space, safety devices, light, and ventilation as facilities handling cylinders.

A properly designed ton-container-handling facility will also have an overhead monorail hoist and motorized trolley of at least 1 800 kg (2 ton) capacity. The monorail should be of sufficient length to allow removal of the container from the delivery vehicle to placement in its storage or use position without being rolled along the ground.

Containers that are not presently in use may be stored on either trunnions or storage cradles. Any container scale and container in use must incorporate trunnions with rollers as part of its design. This requirement is necessary to permit rotating the container so that the valves on the ton container are vertical. Gas feed is from the top valve and liquid feed from the bottom. Holddown chains for each container, both stored and in use, are recommended to prevent container movement, especially in earthquake-prone areas.

Piping and connectors must be kept clear of walkways and work areas. Electrical fixtures should be gas tight and corrosion resistant.

Separate drainage with appropriate seal loops must be provided for each room in the facility.

VAPORIZER FACILITIES. Vaporizers, sometimes referred to as *evaporators,* for chlorine and sulfur dioxide are usually electrically heated water baths that contain pressurized vaporization chambers.

Vaporizers, because they are subject to the buildup of impurities found in liquefied chlorine or sulfur dioxide, must be cleaned periodically. Adequate vertical space must be provided to allow removal of the water bath and chamber from the unit for cleaning. Depending on the type or manufacturer of the vaporizer, it may not be necessary to remove the chamber for cleaning. The frequency of cleaning is a function of the liquid feed rate and chemical quality.

If liquid chlorine or sulfur dioxide feed from ton containers is required, feeding from one container only is recommended. However, if liquid feed is required from containers with manifolds, each container gas valve must also have a manifold to equalize the pressure in the containers. The Chlorine Institute recommends that a gas-pressure equalization manifold be used for liquid connections from ton containers with manifolds (Figure 5.8). Liquid switchover systems allow the continuous supply of liquid feed to the process. These can be used when 100% standby is desired (Figure 5.9).

Liquid piping systems for chlorine or sulfur dioxide should be as short as practical. Most chlorine piping is 19 mm (0.75 in.) or 25 mm (1 in.) in diameter and uses threaded connections and 3 000-lb- (1 400-kg-) rated fittings. The use of schedule 80 seamless carbon steel meets Chlorine Institute requirements. Expansion chambers must be provided on liquid lines that can be isolated by shutoff valves. Appropriate instruments and alarm contacts must be included to provide warning of expansion chamber isolation rupture disc failures.

In the event of a large liquid spill, some form of chlorine or sulfur dioxide neutralization equipment or means for rapid chlorine liquid dispersal should be provided.

CONTAINER HOOKUP. The following precautions should be observed by operators when connecting containers to the gas or liquid pressure manifold:

- If vacuum regulators are not mounted directly on the container for gas feed, a flexible connection is required. This connection may require an isolation valve. If liquid feed is desired, a container-isolation valve is required on each flexible connector.
- Flexible connectors should always be replaced at least annually for safety purposes. Discoloration and crimping are also signs that indicate replacement is necessary. A "crackling" or "screeching" noise when moving the connector indicates metal fatigue and is another signal for replacement.
- The use of a canister-type chlorine gas mask is strongly recommended when changing containers. An alternative is the use of a self-contained breathing apparatus.
- A second person should also be available for assistance in case of emergency. If another individual is not available, then contact should be made with a local public agency (police or fire department) to act as backup.
- All valves must be closed before breaking connections in the pressure lines. Lines left open to the atmosphere will allow moisture to enter the system and can result in corrosion.

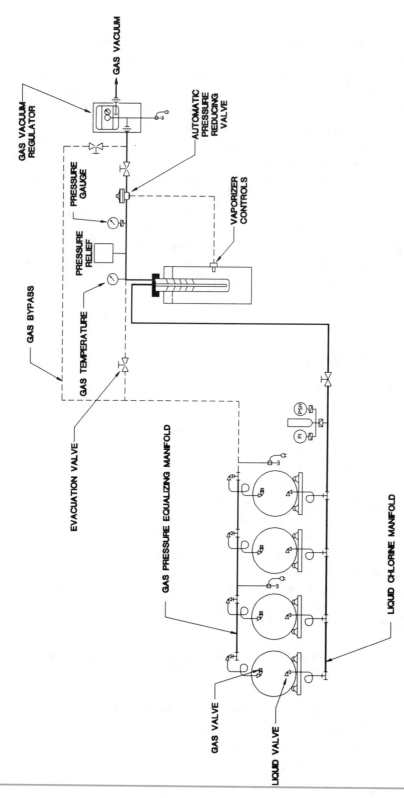

Figure 5.8 Ton container manifold for liquid chlorine withdrawal with optional gas bypass

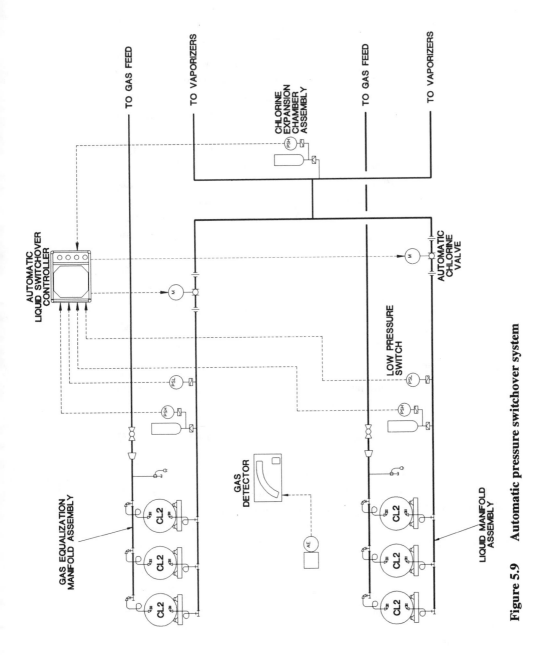

Figure 5.9   Automatic pressure switchover system

- For gas withdrawal from a ton container, the vacuum regulator or flexible connector must always be connected to the vertically aligned top valve. A 70-kg (150-lb) cylinder must be connected when vertical.

ANALYTICAL DETERMINATION OF CHLORINE RESIDUALS

METHODS. Wastewater disinfection requirements have become increasingly more stringent in response to the need for greater protection of water resources and more widespread use of treated wastewater. Furthermore, the potential public health hazards resulting from inadequate or ineffectual treatment have also focused the attention of the water pollution control industry on the need for proper operation and control of disinfection processes. The single most important aspect in controlling the chlorination process is the accurate measurement of chlorine residuals. The techniques described here are for laboratory use. Field operation of a chlorination/dechlorination system is typically accomplished with a continuous amperometric or colorimetric analyzer. Although frequently suggested as a useful control tool, oxidation-reduction potential (ORP) does not measure chlorine residual.

Because analytical results obtained using various methodologies are different, the analytical method used should be specified when reporting results. The reader is referred to the current edition of *Standard Methods for the Examination of Water and Wastewater* (APHA, 1995) for full details on the tests identified here.

Iodometric. The iodometric determination is based on the oxidation-reduction reaction in which iodine is liberated from potassium iodide (KI) solution by chlorine at pH 8 or less (Equation 5.16):

$$HCl + HOCl + 2KI \rightarrow 2KCl + H_2O + I_2 \qquad (5.16)$$

The endpoint is reached at that stage in the titration at which equivalency is obtained. The endpoint may be revealed by changes that can be observed or measured, such as color development or solution-conductance changes. Two ways of determining the endpoint are possible: the starch-iodide method and the amperometric method.

STARCH-IODIDE. In the presence of starch, free iodine produces a blue color indicative of the presence of residual chlorine. Quantitative measurement of total residual chlorine (free and combined) is achieved by titration of released iodine with a standard reducing agent, commonly sodium thiosulfate ($Na_2S_2O_3$), using starch as the indicator (Equation 5.17):

$$I_2 + 2Na_2S_2O_3 \rightarrow Na_2S_4O_6 + 2NaI \qquad (5.17)$$

In wastewater applications, the recommended practice is to use an indirect, backtitration procedure to avoid any contact between the full concentration of the liberated iodine and the wastewater. In this case, standardized iodine solution is used for titration.

AMPEROMETRIC. The amperometric titrator is used to determine the endpoint of a titration that is accomplished by adding a reducing agent, phenylarsine oxide (C_6H_5AsO). Electrical current flow is proportional to the oxidant concentration and decreases as the oxidation-reduction reactions proceed. When these reactions are complete, the current flow will remain nearly constant (the titration endpoint) and will not be decreased by further addition of the reducing agent. This method, when applied as an indirect backtitration procedure as described under starch-iodide, is inherently more accurate than the starch-iodide method. It can be adapted to differentiate quantitatively between free and combined available residual chlorine.

N,N-Diethyl-*p*-Phenylene-Diamine. The N,N-diethyl-*p*-phenylene-diamine (DPD) determination is based on the reaction of free and combined chlorine with DPD to form a red color. In the absence of the iodide ion, only free chlorine will react with the DPD. A subsequent addition of a small amount of iodide acts catalytically to cause monochloramine to produce color quantitatively. The addition of excess iodide evokes a dichloramine (plus nitrogen trichloride) response. An alteration of the reagent addition permits an independent estimate of nitrogen trichloride. Thus, DPD allows differentiation between combined chlorine residuals that can be useful when, for example, controlling a breakpoint chlorination reaction operating at breakpoint would produce minimum combined residuals with an absence of monochloramine. Once color is produced, chlorine is determined in either of two ways: titration or colorimetry.

TITRATION. Ferrous ammonium sulfate (FAS) will decolorize the red DPD solution. Its titer is checked against a potassium dichromate standard solution and calibrated for chlorine. This is a simple and rapid determination that can provide satisfactory analyses with a minimum of laboratory equipment.

COLORIMETRY. A calibration curve for spectrophotometry is prepared from standard chlorine or potassium permanganate ($KMnO_4$) solutions. This is the principle behind packaged, proprietary DPD test kits.

Leuco Crystal Violet. The leuco crystal violet method measures free and total chlorine separately. The compound, N,N-dimethyl aniline, also known by the common name *leuco crystal violet,* reacts instantaneously with free

---

chlorine to form a blue color. Combined residual interference can be avoided by completing the determination within 5 minutes. In the determination of total chlorine residual, free and combined chlorine react first with iodide to produce hypoiodous acid (HOI), which then produces the dye, crystal violet, when mixed with leuco crystal violet. The blue color is stable for days and may be compared to chlorine standards or to a colorimetric standard curve.

Free Available Chlorine Test Syringaldazin. Free available chlorine test syringaldazin, or FACTS, tests for free chlorine in the range of 0.1 to 10 mg/L. The method has been included in this manual to provide an independent method for free chlorine determination. Free chlorine oxidizes syringaldazin (3,5-dimethoxy-4-hydroxy-benzaldine) on a 1:1 molar basis to produce a colored product that is then measured spectrophotometrically. Properly buffered (pH 6.5 to 6.8), this test has been shown to be the most specific colorimetric test for measuring free chlorine.

Note on Orthotolidine. Orthotolidine is no longer an accepted method for residual determination. Despite its availability and its preference by many operators, its use should be discontinued.

SELECTION OF METHOD. Iodometric methods are suitable for measuring chlorine residuals greater than 1 mg/L, although the amperometric endpoint provides greater sensitivity. All acidic iodometric techniques suffer from interferences in proportion to potassium iodide and hydrogen added. The amperometric titration is the standard of comparison for determining chlorine residual because it is little affected by other oxidants, temperature, turbidity, and color. Loss of chlorine can result from stirring and from cleaning and conditioning of electrodes, which is necessary for sharp endpoints. Amperometric titration requires analytical equipment and greater operator skill than do the colorimetric tests.

The DPD tests are operationally simpler than amperometric titration and can differentiate between free and combined residual species. High concentrations of monochloramine interfere with free chlorine determination, as does oxidized manganese. The FAS titration can be performed with a minimum outlay of equipment.

The leuco crystal violet method exhibits less monochloramine interference than does DPD; however, nitrite plus monochloramine, as well as oxidized manganese, can interfere with free chlorine determinations.

Method selection should be determined by the degree of precision required, known or suspected interferences, budget constraints, and the level of operator training.

PROCESS DESIGN REQUIREMENTS

DISINFECTION FACTORS. A major reduction in the level of bacteria in wastewater results from secondary treatment. However, the number of coliform organisms remaining after secondary treatment is high enough to make disinfection necessary to meet National Pollutant Discharge Elimination System requirements for coliform. In general, the more efficiently a WWTP is operated, the easier it will be to disinfect the effluent. Any failure to provide adequate treatment will increase the bacterial level and the chlorine requirement. High solids content increases the chlorine requirement, as does the soluble organic load. For example, care should be exercised in returning digester supernatant liquor to the primary tank. Such supernatant is difficult to stabilize and will increase chlorine demand, should it reach the chlorine contact chamber.

Increases in the proportion of industrial waste in the influent will generally increase the amount of chlorine required for adequate disinfection. Cyanide present in plating waste is a particularly troublesome component. If easily hydrolyzed cyanide is involved, the bacterial enzymes will split the carbon-nitrogen bond, with resultant ammonia formation. Stable complexes are less likely to be hydrolyzed or influence chlorine requirements. In general, when fluctuating percentages of industrial wastes are contained in the influent, difficulty may be anticipated in maintaining a chlorine feed that will ensure adherence to a specified bacterial standard.

Only general guidelines for chlorine dosage can be provided. Approximate chlorine requirements for disinfecting normal domestic wastewater are listed in Table 5.2. The roles of mixing, contacting, and control strategies in maximizing the effectiveness of chlorination are all important factors to consider.

Because the flow of effluent in stabilization ponds or lagoons is placid and often channeled, these treatment processes can produce effluents that create serious problems in the disinfection process. In areas where ice cover is present, effluents may have excessively high chlorine demand because of high concentrations of hydrogen sulfide or other reducing agents.

MIXING. The importance of rapidly mixing chlorine into a wastewater stream to aid in obtaining plug flow has been known for some time. A plug-flow mixing system, in which each part of the cross section is considered to receive equal treatment, achieves approximately a 2-log reduction in the coliform organism concentration over a backmixed system with flow stream short-circuiting. The optimization of mixing chlorine and hypochlorite solutions with wastewater streams is critical whenever significant and reliable reductions in coliforms or other microorganisms are required. Mixing again plays an important part in the process of dechlorination because rapid mixing will aid the reaction completion and help optimize chemical consumption.

Table 5.2 Chlorine design requirements to disinfect normal domestic wastewater as listed in various references

Treatment	Chlorine design dose, mg/L		
	Ten State's Standards (1992)	WPCF (1976)	White (1992)
Prechlorination (mechanically cleaned tanks)	20–25	—	—
Raw wastewater, depending on age	—	6–12 (fresh)	8–15
		12–25 (septic)	15–30
Primary effluent	20	—	8–15
Trickling filter effluent	15	3–10	3–10
Activated sludge effluent	8	2–8	2–8
Sand filter effluent	6	1–5	1–5

The goal of proper mixing is to enhance disinfection by causing the free chlorine in the chlorine solution stream to react as rapidly as possible with the ammonia nitrogen. This reaction is necessary to form monochloramines and avoid prolonged chlorine concentration gradients that may promote the formation of other chlorinated compounds with little or no germicidal efficiency. Ideally, the mixing device should achieve complete mixing of the chlorine and wastewater streams in a fraction of a second.

Another reason for a proper mixing regime is to convert as rapidly as possible the molecular chlorine (Cl_2) in the chlorine solution stream to free chlorine (hypochlorous acid or hypochlorite ion). The chlorine solution discharge from conventional chlorination equipment has a pH less than 2. At this pH, a 2 000-mg/L chlorine solution can contain up to 38% molecular chlorine at atmospheric pressure. In a properly mixed system, the conversion of molecular chlorine to hypochlorous acid will occur in a fraction of a second. This also minimizes the potential for offgasing of chlorine gas at the diffuser in the event the diffuser experiences a negative hydraulic head. Proper mixing, however, will not correct faulty diffuser design.

CLOSED CONDUITS. Proper mixing may be achieved in a closed conduit with turbulent flow by placing a properly designed chlorine diffuser in the center of the cross section of the conduit's flow field. Design considerations include materials of construction, perforation design, and flow velocity.

HYDRAULIC DEVICES. A simple hydraulic jump may be a satisfactory mixing device. The chlorine diffuser, perforated and positioned perpendicular to the water flow, can either be located in the quiet zone upstream of the turbulent zone created by the hydraulic break or directly in the turbulent zone. The disadvantage of locating the diffuser in the turbulent zone is that the zone shifts position when the flow changes. When using a hydraulic jump, the submergence of the diffuser should not be less than 230 mm (9 in.) below the water surface and before the hydraulic jump at minimum flow. The hydraulic jump is usually effective at mixing when the head loss exceeds 0.6 m (2 ft). For adequate mixing, it is also important to use proper engineering practices that include evaluation of the flow characteristics. For example, minimum Reynolds numbers of 1.9×10^4 for pipe flow and Froude numbers between 4.5 and 9 for open channels are recommended (U.S. EPA, 1986).

Because they provide a hydraulic break, Parshall flumes have been used in conjunction with properly designed diffusers to achieve mixing. External mixing devices such as propeller-type mixers may be used.

More recently, proprietary chlorine and sulfur dioxide vacuum-inducing and mixing devices have been developed. Their use during the last several years has been increasing. These devices combine the vacuum-creating capabilities of a high-efficiency venturi with the mixing capability of a high-speed pump. Used in contact chambers, they have provided the mixing necessary for intimate contact and have replaced the venturi ejectors normally provided with chlorination and sulfonation equipment (Figure 5.10). Savings in chemical consumption are claimed. Details can be obtained by contacting the manufacturers.

CONTACTING. Contacting is a separate process from mixing. Both processes are required in an optimized disinfection system, and neither process can substitute for the other. The objective of contacting is to further enhance the inactivation of microorganisms by the disinfection process. This objective is achieved by maintaining intimate contact between the microorganisms in the wastewater stream and a minimum chlorine concentration for a specified period of time.

A chlorine-contacting device may typically take the form of either a pipeline or a serpentine chamber; either device is satisfactory as long as short-circuiting is minimized, plug-flow conditions are closely approached, corners are rounded to reduce dead-flow areas, and the velocity of the contacting stream minimizes solids deposition in the contact chamber or pipeline. Circular chambers tend to demonstrate a significant amount of short-circuiting; therefore, they should not be used for chlorine contacting. The ratio of flow path length to channel width in serpentine tanks should be 50:1 or greater, and the height-to-width ratio of the cross section of the wetted section should not exceed 2:1. Either vertical or horizontal baffling can be used to increase the effective length-to-width ratio of a contacting system; however, care should

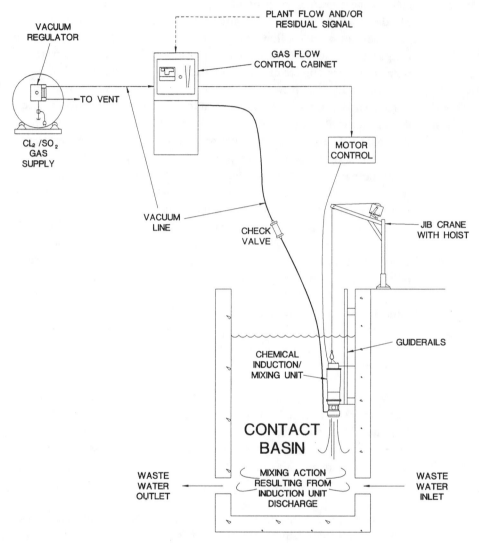

Figure 5.10   Schematic gas induction/mixing device

be exercised to ensure that the placement of the baffles does not create dead-flow areas. The designed length of contact time must meet local regulations. Typically, these regulations call for 30 minutes minimum contact at average flow or 20 minutes at peak flow.

DESIGN AND SELECTION OF EQUIPMENT

CHLORINATORS/SULFONATORS. Chlorine gas feeders are referred to as *chlorinators* to differentiate them from hypochlorinators, which feed a hypochlorite solution. Sulfur dioxide gas feeders are referred to as *sulfonators*.

A chlorinator has the following basic components: a vacuum regulator with an outside vent, feed-rate control, a venturi-operated ejector with check valve, and an indicating flow meter (Figure 5.11). All chlorinator designs in current use consist of these components. Occasional use is made of direct-pressure feeders, but they are rare, particularly in wastewater disinfection. Direct-pressure feeders are used primarily for wastewater disinfection in remote areas or where there is an absence of electrical power preventing the development of a vacuum (Figure 5.12). The vacuum regulator may be found in many different configurations, depending on the manufacturer and the feed-rate capacity of the system. In wastewater disinfection systems, the vacuum regulator is most often located in the chlorine storage room, so that gas leaves the storage room under vacuum. This practice is true regardless of the capacity of the chlorinator and enhances the safety of the installation. The vacuum regulator is a diaphragm-operated device with one side of the diaphragm open to the atmosphere to permit venting, should gas pressure suddenly develop. The other side of the diaphragm is connected to the vacuum source and linked to permit gas flow only under vacuum.

The feed-rate control device may be a simple, manually operated valve or an automatically controlled motor-driven valve. The gas flow meter is calibrated to indicate an instantaneous feed rate in either metric units (grams or kilograms per hour) or English units (pounds per day).

The injector (or ejector) is a vacuum-producing device that consists of a nozzle and throat assembly and a check valve. The device produces a vacuum when a designed water flow passes through a venturi or orifice. The physical principle is known as the *Bernoulli principle*. When water flow ceases, the check valve closes to prevent water from entering the chlorinator. To increase safety in handling chlorine and improve the speed of response to flow or residual changes, a longer gas vacuum line is generally preferred to a long solution line. Therefore, positioning the injector as close as possible to the point of addition should be strongly considered in each installation.

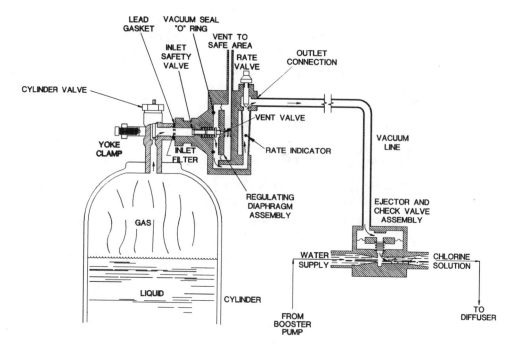

Figure 5.11   Schematic of cylinder-mounted, vacuum-operated gas chlorinator

The injector is a critical component of the chlorinator for two reasons: first, its hydraulic component creates the vacuum under which the system operates; second, it mixes the gas with the makeup water to produce the solution that is injected into the wastewater. If the hydraulic conditions under which the injector must operate are not properly specified, the entire system may operate erratically or not at all. The design of the installation must consider the pressure at the point of injection and the supply pressure needed to the injector. A booster pump designed to provide the correct water flow and pressure must be chosen. The manufacturer of the injector can provide that information.

If addition is into a pipe, the injection fitting consists of a check valve, solution tube, and corporation cock assembly. The solution tube passes through the corporation cock into the pipeline. If the addition is into an open channel, the injector is connected to a diffuser suitably located in the contact chamber. Use of a chemical induction unit negates these devices (Figure 5.10).

CHEMICAL-FEED PUMPS. The chemical-feed pumps used for the feeding of sodium or calcium hypochlorite are referred to as *hypochlorinators*. Like chlorinators, hypochlorinators come in many configurations, but the basic system components are similar in all cases. Sulfite solutions are also fed with chemical-feed pumps. Materials of construction are changed to meet the chemical-handling needs.

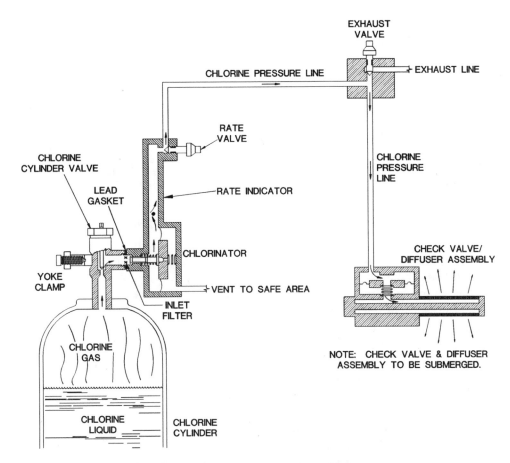

EXHAUST VALVE

CHLORINE PRESSURE LINE

EXHAUST LINE

RATE VALVE

CHLORINE CYLINDER VALVE

CHLORINE PRESSURE LINE

LEAD GASKET

RATE INDICATOR

CHLORINATOR

YOKE CLAMP

CHECK VALVE/ DIFFUSER ASSEMBLY

VENT TO SAFE AREA

INLET FILTER

CHLORINE GAS

NOTE: CHECK VALVE & DIFFUSER ASSEMBLY TO BE SUBMERGED.

CHLORINE LIQUID

CHLORINE CYLINDER

Figure 5.12   Schematic of direct-pressure-operated gas feeder

The basic components are a storage reservoir or mixing tank for the hypochlorite solution; a metering pump that consists of a positive displacement pumping mechanism, motor or solenoid, and feed-rate adjustment device; and an injection device (Figure 5.13). Depending on the size of the system, a plastic or fiber-glass vessel is used to hold a low-strength hypochlorite solution. Hypochlorite solutions are corrosive to metals commonly used in the construction of storage tanks. In addition, if any corrosion products, such as iron, are produced, the hypochlorite solutions will decompose more rapidly. Feeding of calcium hypochlorite will require a mixing device, usually a motorized propeller or agitator located in the tank. Also in the tank is a foot-valve and suction strainer connected to the suction inlet of the hypochlorinator. Mixing takes place in a separate tank.

The hypochlorinator itself can sit either on top of the solution tank on a shelf or on a pump stand. The hypochlorinator consists of a positive displacement diaphragm chamber with inlet and outlet check valves and a drive mechanism. There are several types of drive mechanisms. One variety is a

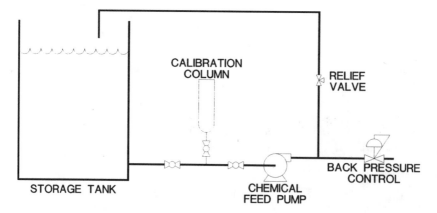

Figure 5.13   Chemical-feed system schematic

constant-speed motor with an adjustment that permits changing the amount of the diaphragm movement by changing the stroke movement. This, in turn, changes the amount of solution feed. Another variety is a variable-speed motor with a similar type of adjustment method. The feed-rate adjustments can be made manually or automatically. The variable-speed motor and automatic stroke length adjustment are controlled by an input signal, either a flow signal (feed-forward), residual signal (feed-back), or both (compound loop). A third type of drive unit consists of a solenoid driven stroke shaft, with independent adjustments for stroke length and frequency of the solenoid operation. This type of drive unit can be operated with manual controls for stroke and frequency or fitted with automatic frequency control, which is analogous to the variable-speed drive on a motor-driven hypochlorinator.

The injection fitting is similar to that used for a gas chlorinator. The amount of solution flow dictates the fitting size.

The hypochlorinator must develop sufficient internal pressure to inject its solution into the contact tank or pipeline.

SULFUR DIOXIDE FEEDERS. Sulfur dioxide gas feeders (sulfonators) are similar in design and construction to chlorinators but composed of different materials. Although some sulfonators are considered by design engineers to be standby chlorinators, this practice is not generally recommended. Misuse by operating personnel can cause inadvertent mixing of the two gases, which could cause an exothermic reaction and lead to equipment failure. Although the sulfonator, usually made of PVC, can be used in an emergency to feed chlorine gas, the chlorinator, usually made of acrylonitrile butadiene styrene (ABS), must not be used to feed sulfur dioxide. Liquid chlorine droplets will soften PVC, while ABS can withstand limited exposure to liquid droplets.

MANIFOLDS AND VACUUM REGULATOR LOCATION. If the vacuum regulator in a chlorination or sulfonation system is not mounted directly on the cylinder, a manifold or pressure piping is required. Gas manifolds consist of a flexible copper connector, usually with an isolation valve(s), and a rigid pipe section of carbon steel with a drip leg (Figure 5.14). If pressure manifolds are in use, reliquefaction must be prevented. Tracing of pressure lines, the use of pressure-reducing valves, insulation of the pressure line, sloping of the line back toward the source, and the use of drip legs at points of direction change and low points in the piping are recommended.

Often, for high-capacity gas manifolds or manifolds that extend for long distances (more than 6 m [20 ft]), a pressure-reducing valve preceded by a gas strainer may be helpful. This requirement is unnecessary if the pressure/vacuum regulator is remotely located as close as possible to the container. Friction losses must be kept to a minimum. The determination of line size should be calculated so that pressure in the line at the gas feeder is at least 4.6 bar (70 psig) to prevent liquefaction. In vacuum systems, the drop should be such that the vacuum at the regulator is greater than 350 mm (14.6 in.) for sonic feed systems and 125 mm (5 in.) for nonsonic systems.

Plant designs should include the use of 100% standby and automatic switching from on-line to standby equipment. The use of vacuum-operated, automatic switchover devices that change from an empty to a full supply of chlorine is recommended (Figure 5.15). These devices are available at operating rates of up to 80 kg/h (4 000 lb/d). When requirements exceed these values, pressure-type switchover systems are used. Because U.S. Department of Transportation regulations do not permit unattended switching of rail tank cars, these devices are not used at tank car installations (U.S. Code, 1994).

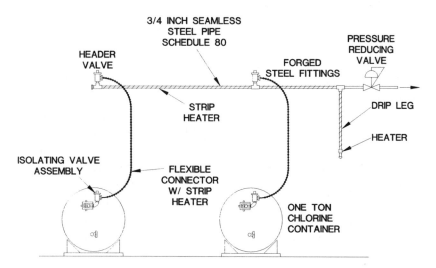

Figure 5.14   Typical gas manifold with heaters

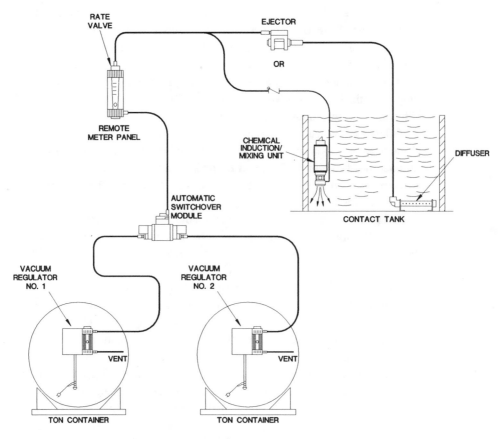

Figure 5.15   Automatic vacuum switchover system

VAPORIZERS. When sulfur dioxide or chlorine is supplied only in tank cars or there is insufficient floor space to permit manifolding of a sufficient number of containers, the liquid chlorine or sulfur dioxide must be converted to gas in a vaporizer. In addition, because gas-withdrawal rates are approximately 8 kg/h (400 lb/d) for chlorine and 4.5 kg/h (225 lb/d) for sulfur dioxide, the desired feed rate must be determined so that the choice of manifolded containers or use of a vaporizer can be decided on during the system design.

A chlorine or sulfur dioxide vaporizer (Figure 5.16) consists of an inner chamber, into which the liquefied gas is introduced, and an outer chamber, which operates as a water bath. The water bath is maintained at a high temperature by immersion heaters. The vaporizer has controls for water level, water temperature, and water bath cathodic protection. The chlorine or sulfur dioxide vapor from the liquid in the inner chamber exits the vaporizer through a superheat baffle, past a rupture-disk-protected relief valve and an automatically controlled pressure-reducing valve, to enter the gas feeder. The use of a gas strainer in the vaporizer outlet piping is recommended.

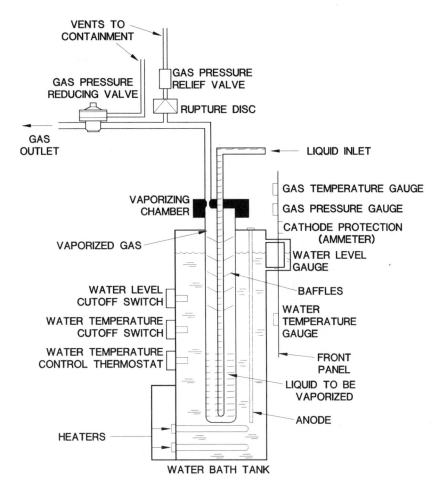

Figure 5.16 Vaporizer schematic

The difference between the operation of vaporizers for sulfur dioxide and chlorine is the vapor pressure of each gas. The vapor pressure of sulfur dioxide is approximately half that of chlorine. Chlorine can be produced at the rate of 8 kg/h (400 lb/d) by natural vaporization within a container. Therefore, systems requiring feed rates of up to 30 to 40 kg/h (1 500 to 2 000 lb/d) capacity may have several containers manifolded and would not require chlorine vaporizers. However, a sulfur dioxide container can produce only 4.5 to 5 kg/h (225 to 250 lb/d) of sulfur dioxide gas. Thus, sulfur dioxide vaporizers are commonly used on systems with capacities as low as 4.5 kg/h (225 lb/d). Alternatively, two or more containers can be manifolded together to achieve the desired gas-flow rate without resorting to the use of a liquid chlorine or sulfur dioxide vaporizer.

RESIDUAL ANALYZERS. One important piece of equipment is an on-line chlorine residual analyzer, which measures and transmits a continuous signal proportional to the chlorine residual in the sample stream. The available types of residual analyzers are amperometric, colorimetric, specific ion, and polarographic probes.

The most commonly used residual analyzer is the amperometric type (Figure 5.17). This analyzer consists of a sample cell with two electrodes, either gold and platinum or gold and copper; buffer feed and reagent (potassium iodide) feed mechanisms; a cell-cleaning mechanism; temperature compensation circuitry; and calibration circuitry. The operation is simple. Two dissimilar metals, in the presence of a chlorine solution, facilitate the passing of a current between them that is proportional to the concentration of chlorine. pH depression to pH 4 stabilizes the signal, and the presence of potassium iodide enables measurement of combined chlorine. This analyzer is used when a total chlorine residual measurement is desired. Generally, the chlorine analysis and recording of a daily quantity of grab samples are required to meet most permit requirements, although the use of a continuous chlorine residual analyzer and recorder also allows a high degree of confidence in the disinfection process to meet the chlorine discharge level regulations of each state. In addition, the use of the continuous analyzer and recorder enables automatic operation and control.

In practice, the accuracy and reliability of an amperometric analyzer depends entirely on the instrument's ability to keep the electrodes in the cell clean or at a constant state of cleanliness, thus stabilizing the signal.

Colorimetric analyzers are not frequently found in WWTPs because wastewater often contains sufficient quantities of suspended solids and turbidity to render an optical cell ineffective. In addition, the DPD method is subjected to interferences in wastewater that cannot always be compensated for by colorimetric analyzers.

Specific ion probes and polarographic probes are somewhat more attractive for determining chlorine residuals in wastewater because they can be located near the point of residual determination and do not generally require a sample pump or a reagent supply. As with amperometric analyzers, the precision and accuracy of residual measurements obtained using probe-type analyzers are influenced by the efficiency of the method designed into them to keep the electrode clean.

The use of ORP devices has also been suggested for residual measurement. However, the reading obtained is not the quantitative measurement of chlorine residual but, rather, a qualitative measurement of ORP. As such, ORP devices cannot be used to measure chlorine residual.

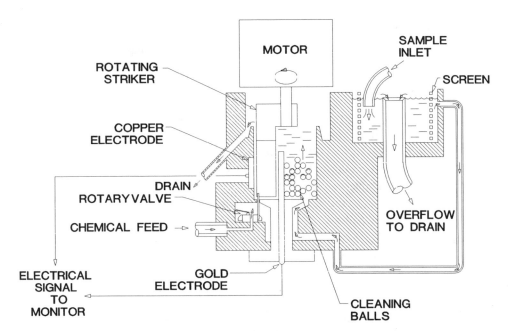

Figure 5.17 **Amperometric analyzer schematic**

MAINTENANCE. An adequate supply of spare parts and the relevant operation and instruction manuals should always be available to the operator and maintenance personnel. Equipment repairs should never be attempted by personnel who are inadequately trained or lack the proper tools or parts.

If standby equipment has not been provided, the operator should devise an adequate standby plan to ensure disinfection in case of an on-line equipment failure.

When servicing equipment, the trained maintenance person should reread the instruction manual before commencing work each time. This review can avoid some of the most common mistakes in repair.

The manufacturer's instructions should always be reviewed, and if questions exist, the manufacturer should be contacted before repair work is begun.

Successful troubleshooting depends on identification of the problem. Begin by speculating as to what the problem might be. Isolate the obvious possibilities: loss of chlorine supply, valves not open, vacuum not created, or vacuum lost because of cracks in lines. If one of these causes is not clearly the problem, the wisest course is to contact the factory service representative. Also, check to see whether all of the feeders are affected. If so, this usually means that the problem is either upstream (gas supply) or downstream (injector hydraulics) and not in the chlorinator at all.

Water can be used to clean most surfaces of chlorinators and sulfonators, although it may be necessary to use dilute muriatic acid on difficult stains and buildups of impurities. Regardless of the cleaning method used, the components must be thoroughly dried (a dry compressed air or a nitrogen supply is recommended) before reassembly. Also before reassembly, components must be checked carefully for organic residues, which should be completely removed.

If, at any time, the operator or maintenance technicians are unsure of what to do, they should immediately stop the repair and ask for assistance from the manufacturer's service department or local service and maintenance personnel. Improperly repaired equipment is especially dangerous to personnel.

The most frequently repaired parts of chlorination or dechlorination systems are the components of the pressure manifold, especially flexible connectors and valves, and the injector and solution system. These two parts of the system should, therefore, be subjected to the most rigorous preventive maintenance possible. Inspection of these parts of the system at least every 6 months is strongly recommended. Replacement of diaphragms and gaskets in injectors every 2 years is also recommended. Replacement of springs will depend on individual manufacturer's requirements.

Any gasketed pressure connection must have its lead gasket replaced each time the joint is broken. Asbestos fiber gaskets are not recommended because they often do not seal properly. Used gaskets must be discarded immediately and never reused.

FEED CONTROL STRATEGIES

There are several ways to control the feed rate of chlorine gas or hypochlorite solutions: manual control, automatic flow proportioning or open-loop control, automatic residual or closed-loop control, or automatic compound-loop control, which combines the flow and residual signals to vary the gas-feed rate. Flow proportioning is sometimes referred to as *feed-forward* control, and residual control is sometimes referred to as *feed-back* control.

Flow pacing is based on the concept that varying the chlorine feed rate in proportion to flow rate will provide adequate quantities of chlorine at any flow. However, in wastewater, this is not always correct. The chlorine demand will vary with flow and can vary independent of flow.

Residual control involves varying the chlorine feed rate based on deviation from a setpoint on a controller. For systems in which the flow rate is nearly constant on a daily basis or is strictly seasonal, this type of control system works well. For systems in which the flow rate varies often and the demand is variable as well, residual control may not be as effective as flow pacing because residual control systems do not react well to large variations in flow rate over short periods of time.

Compound-loop control provides the ability to take a flow and residual input to control gas feed. Flow is the primary drive, while residual is used to trim the gas feed. When the setpoint in a compound-loop control system is further controlled automatically, the configuration is referred to as *cascade control.* Cascade control requires the use of another analyzer downstream of the compound-loop control analyzer. The output from the cascade control analyzer is used to regulate the compound-loop controller.

The choice of control strategy for a particular installation is based on regulatory requirements, existing facilities, wastewater treatment system design, economics, cost effectiveness, and required system maintenance. The more complicated the selected control system, the more likely it is that service requirements will be more exacting and, therefore, will require more training and an increase in the skill level of operating personnel.

MANUAL CONTROL. Manual control of the chlorine feed is the simplest strategy and, because of its simplicity, may often be the most effective method. A manual control system requires considerably less maintenance and operator expertise than any form of automatic control. The basic chlorinator feeds chlorine at a predetermined constant feed rate, which is changed by the operator as required. Manual control has a low capital cost, but it is prone to either overfeeding or underfeeding of chlorine and, therefore, excessive dechlorination, insufficient disinfection, or overchlorination. Manual control systems are used where the flow rate and demand are fairly constant. An example of where this may be appropriate is discharge from lagoon systems.

SEMIAUTOMATIC CONTROL. The same equipment used in manual control systems can often be used to partially automate the operation of a system.

One such system is on–off control. The chlorinator can be turned on and off automatically in response to a signal, such as a wet well level or pump activation. The feeder can be turned on and off by controlling the following: the booster pump, a solenoid valve in the water supply line to the injector or ejector, or a solenoid valve located in the gas vacuum line between the rate-control valve and the injector.

Another option for semiautomatic control is a method known as *band control.* In this technique, two chlorine gas-flow metering tubes are used in the gas vacuum line in conjunction with two vacuum line solenoid valves and a chlorine residual analyzer. The analyzer, or a recorder receiving a signal from the analyzer, is equipped with two setpoint alarm contacts. The contacts are preset at points of maximum and minimum levels of residual. These contacts activate the vacuum line solenoid valves as follows:

- When the residual is lower than the low setpoint, both valves are open;
- When the residual is higher than the high setpoint, both valves are closed; and
- When the residual is between the two setpoints, one valve is closed.

The width of the band and settings on the flow meters should be left to field setting to minimize cycling.

FLOW-PROPORTIONAL CONTROL. In most WWTPs, flow is not discretely variable, nor is it possible or practical to construct equalization basins. Therefore, the control of the chlorination feed is set in proportion to flow, which enables the ratio to be varied by adjusting the dosage. In this strategy, a flow signal is transmitted by a primary flow element to the chlorinator, where an automatic valve opens or closes depending on the signal level (usually 4 to 20 mA dc) from the flow meter (Figure 5.18).

Flow-proportional chlorinators are sized by establishing a design dosage for the chlorine feed rate in grams per hour (pounds per day). This design sets the maximum wastewater flow and maximum chlorine flow dosage at a 1:1 ratio. This means that at a 10% signal from the wastewater flow meter, the chlorinator flow meter will read 10% of full scale; at 90% of the wastewater flow rate, the chlorine flow meter will read 90%; and so on. Flow-proportional controllers have a dosage-control adjustment that allows the operator to vary the design dosage ratio from 5:1 turndown to 1:2 turnup.

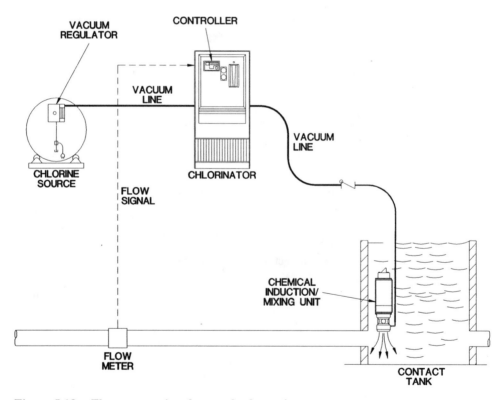

Figure 5.18   Flow-proportional control schematic

In some situations, the use of multiple gas-flow meters in the gas vacuum line, in conjunction with vacuum line solenoid valves, will produce a feed consisting of incremental steps, similar to flow proportioning. The accuracy of this method is limited to the number of metering tubes provided and their settings. Although the chlorine feed tracks the flow variation, the actual chlorine feed is always slightly higher or lower than that of flow-proportional control. A signal from the flow meter to a step-feed controller is needed to activate each step.

RESIDUAL CONTROL. One of the criteria for successful design of residual control systems (Figure 5.19) is to minimize lag time. Lag time consists of four major components: first, the time required for the flow to pass from the injection point to the sampling point; second, the time required for the flow to pass from the sample point to the analyzer (including the speed of response of the analyzer); third, the time required for the chlorine gas or chlorine solution to reach the diffuser; and fourth, the response speed of the control valve in the chlorinator.

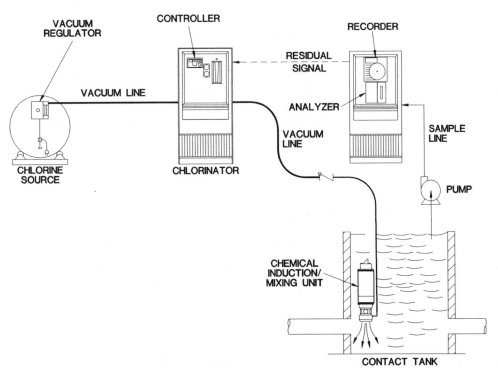

Figure 5.19   Residual control schematic

The first component of lag time depends on the flow rate and the distance between the injection point and the sample point. To limit the lag time, this distance should be minimized. The chlorination reaction in wastewater requires a period of time for completion. Therefore, to obtain an accurate residual, the sample point cannot be located near the injection point. The disinfection pathway is initiated immediately after the addition of the chlorine to the wastewater if proper mixing occurs. Thus, for optimal control, the sample point should be located at a distance that corresponds to approximately a 90-second travel time at maximum flow rate from the injection point.

The distance from the sample point to the analyzer and speed of response of the analyzer can be minimized as well by locating the analyzer as close as possible to the contact chamber and the sample point. Locating the analyzer in the control room or some similar location contributes to poor control. The use of a dedicated analyzer produces a consistent, rapid reading because recalibration is unnecessary and sample line cleaning is not required. The sample line should include the capability of periodic cleaning with high concentrations of chlorine to remove any buildups of algae and slime.

When the injector and control valve are located at the diffuser site, any change in chlorine feed called for by the control signal is sensed rapidly because the distance to the point of addition is minimized.

Finally, the speed of response of the control valve in the chlorinator is relatively insignificant compared to the other three factors.

This discussion does not suggest that the chlorination reaction takes 90 seconds or that contact for a minimum of 30 minutes is unnecessary. Rather, the implication is that after 90 seconds, the chlorination reactions are sufficiently complete to be measured for control purposes, and if the control system is functioning properly, the chlorine residual after 30 minutes of contact may be obtained by manual sampling and analysis.

If manual sampling is insufficient to provide an adequate safeguard against improper disinfection, an additional dedicated residual analyzer can be installed at a sample point at least 30 minutes detention downstream of the injection point. The rationale for using dedicated analyzers in lieu of sample switching arrangements is the following: if an analyzer is alternately reading the control sample (90 seconds) for 45 minutes every hour and the contact chamber effluent sample (minimum 30 minutes detention), it cannot be used for control during the 15 minutes of every hour that it is reading the end-of-chamber sample. Given the time required to flush the analyzer and stabilize the reading every time the sample switching occurs, an analyzer in this type of system probably is usable as a valid control signal source for 30 to 35 minutes per hour and produces 8 to 10 minutes of valid sample data from the end of the contact chamber every hour. In practice, the use of dedicated analyzers allows continuous measurement and control, and the additional cost is negligible compared to the benefit derived.

COMPOUND-LOOP CONTROL. Because chlorine demand is not exactly proportional to flow rate, and because it is often impossible to regulate flow rate to the extent that a manual control or residual control system becomes practical, many WWTPs are equipped with a chlorination system that combines the advantage of gross regulation of chlorine feed using flow proportioning with adjustment of the flow proportion (dosage) (Figure 5.20). These compound-loop systems are subject to the same cautions with respect to flow meter sizing and lag time as flow pacing and residual control systems.

A compound-loop system consists of two interlocking control loops: the flow loop and the residual loop. These loops can be interlocked in one of several ways. One way is to have the signal from a residual controller modulate the dosage adjustment of the chlorinator. Alternatively, the signals from the analyzer and the flow meter can be sent to a multiplier, which then sends a composite signal corresponding to mass flow to the chlorinator. The chlorinator then injects the appropriate amount of chlorine based on this mass signal.

An advantage of the compound-loop controller is that it can introduce damping into the system, enabling the system to avoid uncontrollable oscillation that results from excess lag time. The multiplier type of compound-loop

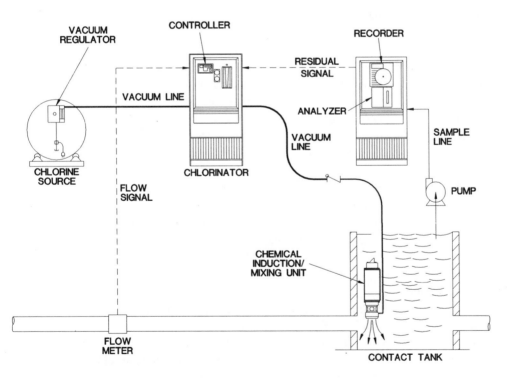

Figure 5.20   Compound-loop schematic

controller has the advantage of being able to react quickly to rapid changes in residual (and therefore demand). The choice of controller for a multiplier-type loop system should be based on the specific requirements of installation.

In WWTPs, a compound-loop system not only provides more accurate control, but, equally important, it can save costs by minimizing the overfeeding of chlorine by operators and reducing the amount of chlorine and chlorine byproducts in the effluent.

CASCADE CONTROL. In cascade control systems, additional instrumentation, measurement, and input to the control scheme are used. In this system, an additional chlorine residual analyzer is provided to sample the contact chamber (Figure 5.21). Located at or near the contact chamber discharge, this second analyzer provides input to the control scheme by varying the setpoint of the compound-loop analyzer, thereby providing an additional damping effect on the variation of chlorine residual.

MANUAL INJECTION POINTS. Often, multiple chlorine injection points are provided in the WWTP, for example, at the headworks or clarifier (primary or secondary) or in the return sludge line. These injection points are used primarily for odor or biological control and are manually controlled to meet specific operating problems. Automatic control is not used at these application points.

Historically, solution redirection systems using manual valves and glass flow meters for indicating chlorine solution flow have been used to accomplish this task. For several reasons, the use of manual valves and glass flow meters in chlorine solution control is not recommended. First, the use of a manual control station subsequent to the automatic control valve unnecessarily complicates the control loop and can render the chlorination control ineffective. Second, some wastewater releases dissolved gases, as well as chlorine, when used as solution makeup water. This will affect the accuracy of the flow meter reading. The use of such a solution redirection system implies that the injectors are located inside the chlorine control room. To minimize lag time in automatic control systems, it is recommended that the injector be located as close as possible to the injection point and that the solution line be as short as possible. This also decreases the chlorine breakout caused by introducing a moisture-saturated gas to the mixing chamber and, thus, reduces chlorine consumption.

The recommended method for providing manual injection points is to use manual gas-metering stations with injectors located as close to the actual injection point as possible (Figure 5.22). Both the injectors and the metering stations may be located directly at the application point instead of in the chlorinator control room. Alternately, the use of a second chlorinator, manually fed, permits feed to other points and does not influence the automatic control of the chlorinator.

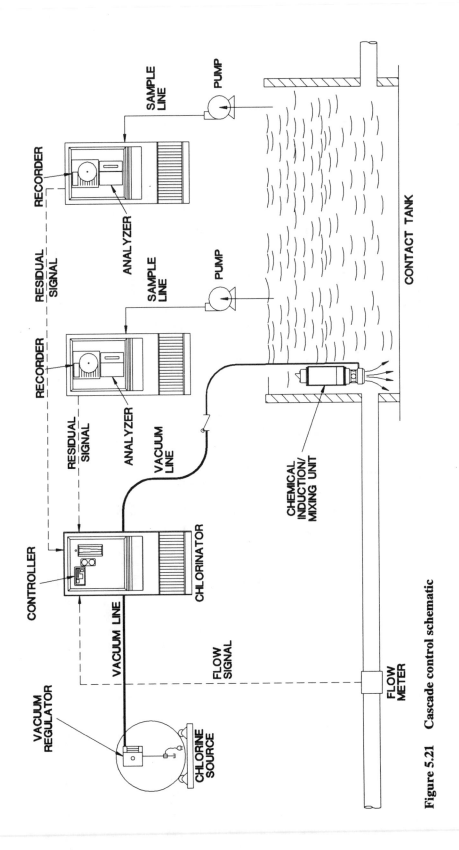

Figure 5.21   Cascade control schematic

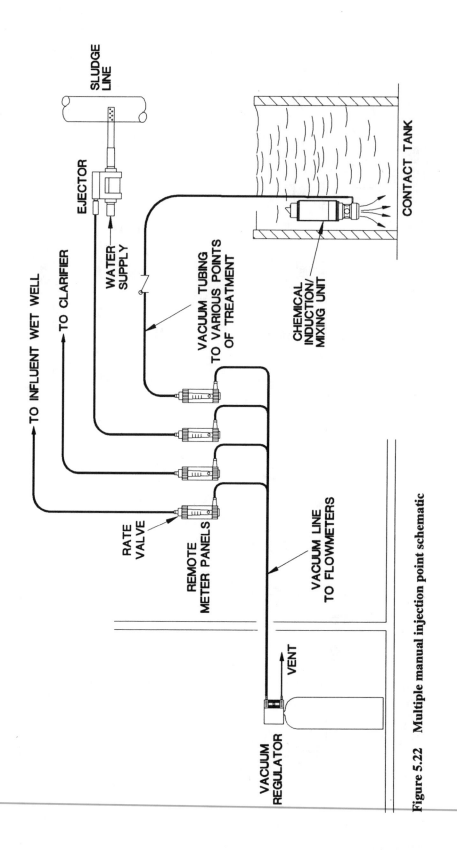

Figure 5.22   Multiple manual injection point schematic

Wastewater Disinfection

DECHLORINATION CONTROL. In general, the same principles identified for chlorination systems also apply to dechlorination systems. The continuous measurement of sulfite residuals can be accomplished either directly or indirectly. The commonly accepted practice is to use a chlorine residual analyzer and shift the zero point by adding a known amount of oxidant (chlorine). This enables a residual of either chlorine or sulfur dioxide to be determined and used in the control scheme. Alternately, the Renton system, which is a variation of the zero-shifted analyzer, has been used successfully in some areas (Finger *et al.*, 1983).

Two types of control systems for sulfonation are often used. In WWTPs that are not required to completely dechlorinate their effluents, it is possible to use a feed-back control system (Figure 5.23), whereby the analyzer measures the chlorine residual a short time after the injection and mixing of the sulfur dioxide. Lag time between the injection point and sample point is minimal because the dechlorination reaction is almost instantaneous. The setpoint signal is used as an inverse controller: as the chlorine residual increases, the feed rate of sulfur dioxide also increases.

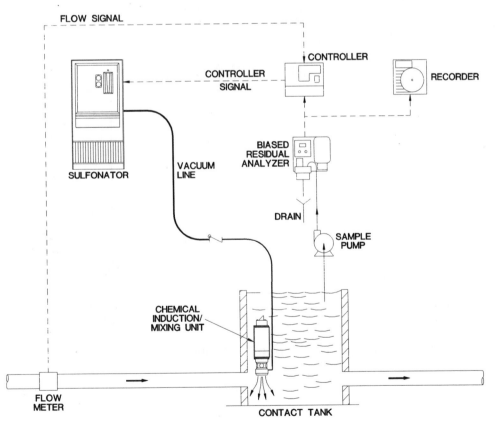

Figure 5.23   Dechlorination feed-back schematic

In those plants that must completely dechlorinate their effluent and do not have either a biased or direct reading analyzer, the feed-back control system may not be practical. Plants of this type can use a feed-forward system, with a multiplier to send a mass flow signal to the sulfonator based on the signal from the analyzer located at the end of the contact chamber (Figure 5.24). A modification of the standard feed-forward design, in which a gas-flow transmitter is installed in the vacuum line of the sulfonator, can also be used. The transmitter measures the exact flow of sulfur dioxide through the feeder and transmits this signal to a ratio controller. The controller compares the multiplied signal, in kg/h (lb/d) of chlorine in the water, with the measured feed rate of sulfur dioxide, and provides a control signal output that is proportional to the ratio required between the two gases for proper dechlorination.

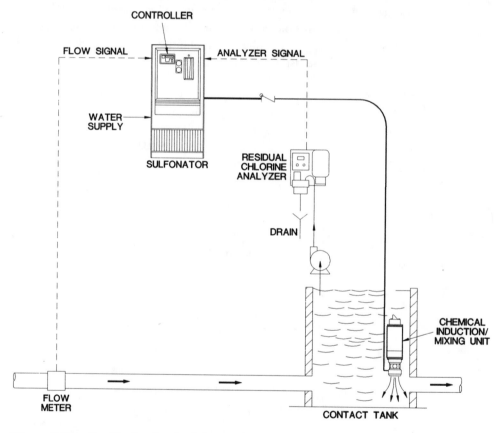

Figure 5.24   Dechlorination feed-forward schematic

REFERENCES

American Public Health Association (1995) *Standard Methods for the Examination of Water and Wastewater.* 19th Ed., Washington, D.C.

American Water Works Association (1973) *Water Chlorination Principles and Practices.* Manual M-20, Denver, Colo.

Boliden Intertrade (1979) *Sulfur Dioxide Technical Handbook.* Atlanta, Ga.

Building Officials and Code Administration International, Inc. (1993) *National Fire Code.* Country Club Hills, Ill.

Chlorine Institute (1986) *The Chlorine Manual.* 5th Ed., Washington, D.C.

Clarke, N.A., *et al.* (1964) *Human Enteric Viruses in Water: Source, Survival and Removability, Advances in Water Pollution Research.* Vol. 2, Pergamon Press, London, U.K., 523.

Compressed Gas Association (1988) *Sulfur Dioxide.* 4th Ed., Pamphlet G-3, Arlington, Va.

Finger, R.E., *et al.* (1983) Development of an On-Line Zero Chlorine Residual Measurement and Control System. Paper presented at the 56th Annu. Conf. Water Pollut. Control Fed., Atlanta, Ga.

Jafvert, C.T., and Valentine, R.L. (1992) Reaction Scheme for the Chlorination of Ammoniacal Water. *Environ. Sci. Technol., **26,** 577.

Southern Building Congress International (1991) *Standard Fire Prevention Code.* Birmingham, Ala.

Ten States Standards (1992) *Standards for Sewage Works.* Upper Miss. River Board of Public Health; Great Lakes Board of Health, Albany, N.Y.

Uniform Fire Code (1994). Int. Fire Code Inst., Austin, Tex.; Int. Conf. Build. Off., Whittier, Calif.; West. Fire Chiefs Assoc., Temecula, Calif.

U.S. Code of Federal Regulations (1993) Title 29, 29 CFR, 1910.

U.S. Code of Federal Regulations (1994) Title 29, 29 CFR 174.67.

U.S. Environmental Protection Agency (1986) *Design Manual Municipal Wastewater Disinfection.* EPA-625/1-86-021, Office Res. Dev., Cincinnati, Ohio.

Vulcan Chemicals (1993) *Sodium Chlorite Manual.* Birmingham, Ala.

Water Pollution Control Federation (1976) *Chlorination.* Manual of Practice No. 4, Washington, D.C.

White, G.C. (1992) *Handbook of Chlorination and Alternative Disinfectants.* 3rd Ed., Van Nostrand Reinhold, New York, N.Y.

Chapter 6
Ozone Disinfection

158 History
159 Chemistry
159   Physical and Chemical Properties
161   Ozone Chemistry in Aqueous Solution
161   Ozone Decomposition
161   Reaction Pathways and the Fate of Ozone
165   Measuring Ozone Concentration
165   Gas-Phase Concentration Measurement
167   Aqueous-Phase Concentration Measurement
168   Toxicological Properties
168   Occupational Exposure
170   Residual Toxicity in the Water Environment
172 Disinfection
172   Mechanisms
173   Factors Affecting Disinfection Efficacy
175   Effect on Bacteria
175   Effect on Viruses
177   Effect on Protozoa
179 Design
179   Feasibility
180   Bench-Scale Tests
180   Pilot-Scale Tests
181   Process Components
181   Feed Gas Selection and Preparation
181   Source Gas

182 Gas Treatment Systems
184 Ozone Generation
184   Theory
185   Types of Generators
186   Cooling
188 Contacting
188   Objectives
189   Gas Mass Transfer
190   Disinfection Efficiency
191   Types of Contactors
195   Contactor Design Considerations
195   Fine-Bubble Diffusers
197   Injector and Static Mixer
197   Turbines
197   Process Control
198 Offgas Destruction
198 Materials
198 Process Design Example
202   Step 1: Determine the Initial Ozone Demand and Slope of the Dose-Response Curve
203   Step 2: Choose the Ozone-Transfer Efficiency and Size the Contactor
203   Step 3: Determine the Design Ozone Production Rate
203   Step 4: Select the Number of Ozone Generators
205   Step 5: Determine the Size and Number of Air Compressors
206   Step 6: Determine the Size and Number of Desiccant Dryers

208 **Operation, Maintenance, and Safety**   211   Maintenance
208   Operation   211   General
208   General   211   Gas Preparation
209   Gas Preparation   212   Electrical Power Supply
210   Electrical Power Supply   212   Ozone Generator
210   Ozone Generator   212   Ozone Contactor
210   Ozone Contactor   213   Associated Equipment
210   Offgas-Destruction System   213   Safety
210   Additional Considerations   215 References

HISTORY

Disinfection of wastewater effluents has been extensively studied in the U.S. since the early 1970s in response to the inclusion of a universal fecal coliform standard for wastewater discharges in the Clean Water Act of 1972. The potentially adverse effects caused by the extensive use of chlorine as a wastewater disinfectant were recognized by the Canadian and U.S. governments and led to a discontinuation of the universal disinfection requirement in the U.S. in 1976. The search for effective alternatives to chlorine led to the evaluation of other halogens, ultraviolet radiation, and ozone as technically feasible processes. Ultraviolet (UV) radiation and ozone have been considered the most feasible alternatives to chlorination. Both have demonstrated effective reduction of pathogenic organisms in water and wastewater environments. The U.S. Environmental Protection Agency (U.S. EPA) sponsored research during the 1970s and early 1980s to help utilities meet the requirements in a cost- effective fashion. The culminating document sponsored by U.S. EPA was a wastewater disinfection design manual (Stover *et al.*, 1986).

Ozone can be used for a variety of purposes including odor control, chemical oxidation, and disinfection. The advantages of ozonation systems include the following (Rosen, 1976):

- Few safety problems associated with transportation and storage;
- Excellent virucidal and bactericidal properties;
- Shorter feasible treatment times (approximately 10 minutes for ozone compared to 30 to 45 minutes with chlorine);
- Less observed dependency on pH and temperature;
- Improvement of receiving stream quality resulting from high dissolved oxygen concentration of ozone-disinfected effluent;
- No increase of total dissolved solids in ozone-treated water;
- Wastewater quality improvements (including effluent color, odor, and turbidity reductions); and
- Easy extension of ozone to tertiary polishing.

Ozone disinfection has an important feature in common with an activated-sludge facility: they both use oxygen as a fundamental part of the process. When ozone disinfection is used in conjunction with a pure-oxygen activated-sludge system, the ozone contact chamber offgases may be recycled back to the aeration tanks after passing through an ozone decomposition device (Rosen, 1973). This results in more efficient use of the oxygen because the ozone not absorbed in the contactor may be decomposed to oxygen and used in the aeration tanks. Oxygen must be the gas used for ozone generation to take advantage of this recycle benefit.

Authorization of the Safe Drinking Water Act in 1986 led to increased interest in water disinfection using ozone because of its efficacy against viruses, bacteria, and protozoa, and the reduction in disinfection byproducts. The benefits from the renewed interest in ozone include improvements in ozone equipment design. In particular, the efficiency of ozone generation and gas-transfer systems has improved greatly since the late 1980s.

CHEMISTRY

PHYSICAL AND CHEMICAL PROPERTIES. Ozone, a triatomic allotrope of oxygen, is ubiquitous; it occurs in nature and can be formed from oxygen but, in either case, is always in dilute form. Ozone has a pungent odor described as resembling freshly mown hay, and at a concentration of approximately 0.01 ppm (v/v), easily detectable in air by most people. The odor is particularly evident following electrical storms or similar discharges. It is of interest to note that everyday objects such as photocopiers and laser printers produce ozone at detectable levels.

Ozone gas is reported to be distinctly blue in hue, and it condenses to a dark blue liquid below –111.9°C (Hann and Manley, 1952). Liquid ozone will readily explode at approximately 30% (w/w), as will compressed gaseous concentrations in air–ozone mixtures. This detonation property of compressed ozone requires that ozone be generated on site rather than at a central facility. Ozone displays a strong oxidizing ability in water but becomes increasingly unstable as trace reducing agents are introduced. Pure ozone is 12.5 times more soluble in water than is oxygen. However, the low partial pressure of ozone in the carrier gas from commercial generators (1 to 10% w/w) makes effective mass transfer to the liquid phase essential for economical operation of an ozone disinfection system. When ozone-reactive materials are present in water, the ozone mass-transfer efficiency is increased. However, disinfection under conditions of high ozone demand can be difficult depending on the relative reaction rates of the components of the wastewater. Each wastewater stream will be different. The physical and chemical properties of ozone are summarized in Table 6.1.

Table 6.1   Selected properties of pure ozone (Weast, 1987)

Property	Value
Molecular weight	48
Boiling point	−111.9°C
Freezing point	−192.5°C
Critical temperature	−12.1°C
Critical pressure	5.53 MPa (54.6 atm)
Density (liquid)	1.358 g/mL at −112°C 1.571 g/mL at −183°C 1.614 g/mL at −195.4°C
Density (gas)	2.144 g/L at 0°C, 1 atm
Heat capacity (gas)	33.3 J/g·mol°C (7.95 cal/g·mol°C) at −173°C
Heat of vaporization	15.19 kJ/mol (3 629 cal/mol) at −112°C
Viscosity (liquid)	1.57 mPa·s (1.57 centipoise) at −183°C
Surface tension	38.1 mN/m (38.1 dyne/cm) at −182.7°C

The concentration of ozone in the gas phase can be reported in a variety of units, including parts per million by volume (ppm v/v), milligrams per standard litre of gas (mg/L$_{STP}$), or percent weight (% w/w). Conversions between the various units are facilitated by the equations in Table 6.2. Gas-phase concentrations from generators and in the offgas of reactors are typically expressed in %O_3 (w/w). Ozone residual, transferred dose, and utilized ozone are expressed in mg O_3/L of water. Knowledge of the gas-flow rate in and out of the reactor and ozone concentration in the gas phase permits a mass balance of ozone to be calculated for the ozone contactor.

Table 6.2   Ozone concentration relationships at standard temperature and pressure (standard temperature and pressure are 273.15°K and 101.33 kPa)

To convert from	To	Calculation
w/w%	Concentration, g/L	$\dfrac{1\,429 \times O_3\ \text{w/w\%}}{100 - (0.333 \times O_3\ \text{w/w\%})}$
Concentration (g/L)	v/v%	$0.046\,666\,7 \times C\ (\text{g/L})$
v/v%	ppm-v	$1.0 \times 10^{-4} \times (O_3\ \text{v/v\%})$
w/w%	v/v%	$1.492\,1 \times (O_3\ \text{w/w\%})$

OZONE CHEMISTRY IN AQUEOUS SOLUTION. To understand the inactivation of microorganisms using ozone, one must first understand the chemistry of ozone in aqueous solution, particularly in the presence of competing reactions such as the oxidation of reduced inorganic species and organic compounds.

Ozone Decomposition. Two mechanisms for ozone decomposition have been proposed: the Staehelin, Bader, and Hoigné (SBH) mechanism and the Tomiyasu, Fukutomi, and Gordon (TFG) mechanism (Chelkowska *et al.,* 1992; Staehelin and Hoigné, 1981 and 1985; and Tomiyasu *et al.,* 1985). Both schemes describe ozone decomposition in terms of three distinct phases, although the specific compounds and reactions vary (Figure 6.1). These phases are as follows (Langlais *et al.,* 1991):

1. Initiation,
2. Propagation or promotion, and
3. Inhibition.

Initiation is the rate-determining step in ozone decomposition. This stage involves a reaction between ozone and an initiator to form a superoxide radical ion (O_2^-). The O_2^- ion then reacts with ozone, causing further ozone decomposition. A variety of agents such as hydroxide ions, ultraviolet radiation, and humic substances can act as initiators.

Hydroxyl radicals (OH^-) are among the byproducts formed during ozone decomposition. Promoters are substances capable of regenerating O_2^- ions from hydroxyl radicals. Hence, ozone decomposition will continue as long as there are sufficient concentrations of promoters present in the solution. Inhibition occurs when hydroxyl radicals are consumed without regenerating O_2^-. Carbonate and bicarbonate ions (wastewater alkalinity) are two common inhibitors. A high concentration of inhibitors in the wastewater can totally inhibit the free-radical chain reactions, reducing the ozone decomposition rate (Doré *et al.,* 1987, and Hoigné, 1982). Langlais *et al.* (1991) can be consulted for specific details of the compounds and reactions involved in the two decomposition mechanisms.

Ozone decomposition is influenced by pH, temperature, UV radiation, the concentration of ozone, and the presence of scavengers such as carbonate and bicarbonate (Tomiyasu *et al.,* 1985). For example, ozone decomposition increases at higher temperatures.

Reaction Pathways and the Fate of Ozone. From an understanding of ozone decomposition chemistry, two main reaction schemes have been proposed for ozone in water: a direct reaction by ozone itself and a reaction involving the radical products of ozone decomposition (Hoigné and Bader, 1975). The first mechanism, the direct oxidation of compounds by the ozone molecule, involves reactions that are selective and can take several minutes.

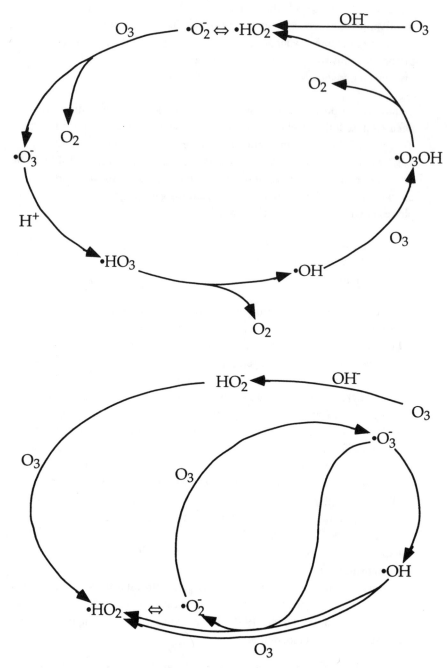

Figure 6.1   Two mechanisms for ozone decomposition in pure water
(top, Staehelin, Bader, and Hoigné model; bottom,
Tomiyasu, Fukutomi, and Gordon model) (Helmer, 1992)

Therefore, in the absence of ozone decomposition, the ozone concentration may remain relatively constant over short time intervals (measured in minutes). The second mechanism is through the oxidation of compounds by ozone decomposition products, principally believed to be the hydroxyl radical. This radical is highly reactive and has a life span of only a few microseconds in water.

The predominant reaction pathway will depend on the characteristics of the wastewater. High concentrations of initiator substances will promote the radical route, leading to rapid consumption ozone, whereas high concentrations of scavenger species will promote the direct route.

The fate of ozone and organics in wastewater is presented in Figure 6.2. Ozone may be physically stripped, may react directly with organics to form an oxidized product, or may enter into a series of radical reactions. In the radical pathway, ozone is converted to the hydroxyl radical by reacting with the

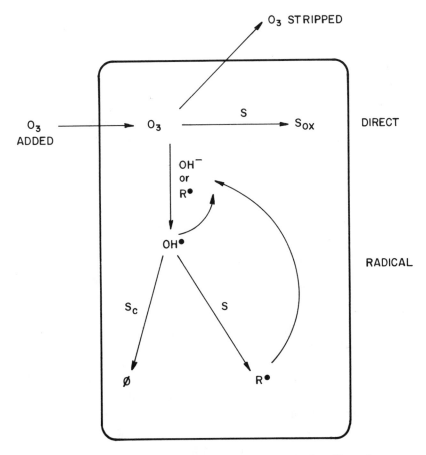

Figure 6.2   **Typical ozone reactions in aqueous solution (S = substrate, R = radical, S_c = scavenger, and ϕ = activated oxidant) (Hoigné, 1982)**

hydroxide ion or another free radical, after which it reacts either with the organics or with extraneous material (scavengers) such as the carbonate ion (Chelkowska *et al.*, 1992).

The byproducts from the reaction of ozone with organic materials in water have been identified (Glaze, 1986; Guittonneau *et al.*, 1992; and Trussell, 1992). In general, the reaction of ozone with organic materials leads to the destruction of the original molecule, often forming a more polar biodegradable product with a lower molecular weight (Huck *et al.*, 1992; Langlais *et al.*, 1989; and Somich *et al.*, 1990). However, in some cases, particularly those involving pesticides, a more toxic or virulent intermediary may be formed (Maggiolo, 1978). Furthermore, such formations as hydroxyhydroperoxide, peroxide, and ozonide have been predicted by chemical-reaction models applied to ozonolysis of specific carbon–carbon double bonds (Maggiolo, 1978). These long-life intermediate species may be of significance to human health and of potential importance in determining the environmental consequences of ozone used as a wastewater disinfectant. However, more research into the general byproducts of wastewater ozonation is needed.

Because of its high oxidation potential, the inorganic chemistry of ozone involves many other elements. While ozone will not react with metal ions that exhibit one oxidation state, transition metals are oxidized to a stable, high oxidation state. Ozone will oxidize ferrous iron to ferric; arsenite to arsenate; and manganese to manganate, manganese dioxide, or permanganate (Nebel, 1981). All members of the halide series, except fluorine, react with ozone. Bromide can be oxidized to bromate through several intermediary steps, during which bromoform may also be produced (ozonation of sea water) (Chrostowski *et al.*, 1982). Because of the adverse health effects associated with bromate formation on ozonation of bromide in drinking water, significantly more data are available on the chemistry of ozone, bromide, and bromate (Krasner *et al.*, 1993; Singer, 1994; and Symons *et al.*, 1994). Other research has shown that bromoform is not produced when drinking water containing bromide and/or bromate (and a humic acid precursor) is ozonated (Graham *et al.*, 1992). Although reaction with chloride is limited by slow kinetics, the reaction of ozone with iodide forms the basis for several analytical techniques used in ozone residual detection.

Most of the inorganic reaction chemistry of ozone has been studied in relation to direct ozone reaction processes. Oxidation of many heavy metals by ozone leaves them in their least soluble form, facilitating removal by sedimentation and/or filtration. Ozone oxidation of cyanide to cyanate has been used to treat wastewater produced by metal-plating and paint-stripping industries. Oxidation of the ammonium ion depends on the pH of the stream, which controls the overall reaction rate. Only under high-pH conditions can ozone oxidize ammonium to nitrate and various intermediate species (Corless *et al.*, 1989; Hoigné and Bader, 1978; and Singer and Zilli, 1975).

The presence of organic matter inhibits the oxidation of inorganics by ozonation for three reasons (Langlais *et al.,* 1991):

- The organics may act as promoters or inhibitors, thus changing the reaction pathway;
- The organics may complex with metals, thus preventing their removal; and
- The organics compete with the inorganics for oxidant.

Likewise, during disinfection, both organic and inorganic matter will compete with microorganisms for oxidant. This is known as *ozone demand*: higher dosages of ozone is required to achieve disinfection because of the competing reactions that take place preferentially over the disinfection reactions.

A series of papers has been published by Hoigné and coworkers investigating the reaction rate constants of various organic and inorganic compounds with ozone in water (Hoigné and Bader, 1983a and 1983b; and Hoigné *et al.,* 1985). An exhaustive list of the rate constants of organic and inorganic compounds and ozone is documented in these papers.

Another broad area of ozone reactions with inorganic species involves the chemistry of colloidal suspensions and enhanced coagulation when using ozone as an oxidant (Chang and Singer, 1991; Chheda *et al.,* 1992; Edwards and Benjamin, 1991; Jekel, 1994; and Maier, 1984). During the design stage of an ozone disinfection system, enhanced particle removal may be considered a secondary benefit that enhances the economic viability of the process (Rakness *et al.,* 1992).

MEASURING OZONE CONCENTRATION. Gas-Phase Concentration Measurement. Concentration measurements of ozone in the gas phase are required for two purposes: evaluation of process gas that is applied to the contactors and process control through offgas monitoring. The recommended analytical techniques for determining ozone in the gas phase include the following (Gordon *et al.,* 1992):

- Ultraviolet absorbance,
- Chemiluminescence,
- Calorimetry, and
- The iodometric method.

For details on how to perform these procedures, the reader should refer to Gordon *et al.* (1992), *Standard Methods for the Examination of Water and Wastewater* (APHA, 1995), or various publications from the International Ozone Association (Birdsall *et al.,* 1952, and International Ozone Association, 1987f and 1987g).

Ozone absorbs light in the short UV wavelength region with a maximum absorption at 253.7 nm, providing a convenient means of measuring ozone

(Gordon *et al.,* 1992). A number of commercially available products can measure the absorbance of a process gas stream with and without ozone. The difference in absorbance can be attributed to ozone, and the concentration is calculated from the molar absorption coefficient and the light path length. Errors in measurement can arise if the reference gas stream (no ozone) is not treated identically to the process gas stream (ozone). This is particularly true when analyzing offgas ozone concentrations in which the process gas has become saturated with water and has stripped volatile organics from the water column.

The chemiluminescence method involves the reaction between ethylene and ozone, which produces formaldehyde in an excited state. The formaldehyde emits photons in proportion to the ozone concentration. The photons are then measured spectrophotometrically. This procedure has been adopted by U.S. EPA as its reference method for determining low concentrations of ozone in the ambient atmosphere (Gordon *et al.,* 1992). The method is calibrated using the direct UV absorption procedure. The chemiluminescence method is used to monitor ozone in the ambient atmosphere and, hence, has application as an occupational health device in ozone-generating facilities.

The enthalpy of decomposition of ozone ($\Delta H = 144.41$ kJ/mole) is the basis for the calorimetry procedure. Thermocouples measure the temperature difference between the inlet gas and the gas within the measuring cell, which is then correlated to the ozone concentration (Gordon *et al.,* 1992). This method has been used to measure process gas concentrations of ozone and has been demonstrated to improve in precision as the concentration of ozone increases. Interferences have not been reported.

Iodometry is based on the oxidation of iodide ion to iodine through the following reaction:

$$O_3 + 3I^- + H_2O \rightarrow O_2 + I_3^- + 2OH^- \qquad (6.1)$$

Ozone must first be absorbed into a potassium iodide solution, with the gas volume measured using a calibrated wet test meter. The mass of ozone that reacts with the iodide is determined by titration to a starch endpoint. Several variations of this method, using different buffers and indicators, have been documented (Gordon *et al.,* 1992).

For many years, the iodometric method was used as the standard to which all other analytical procedures were compared (Hamon, 1982). This method is based on an assumed stoichiometry of 1 mole of iodine liberated for each mole of ozone. However, iodide ion stoichiometry has been found to range from 0.65 to 1.5 moles of iodine per mole of ozone under various conditions (Grunwell *et al.,* 1986). The pH, buffer composition and concentration, iodide ion concentration, sampling technique, reaction time, and presence or absence of any material that can oxidize iodide ion can all interfere with the results (Gordon *et al.,* 1992). For these reasons, the iodometric method is recommended only as a check for other methods or as a relative measure of ozone for control purposes (Gordon *et al.*, 1992, and Langlais *et al.*, 1991).

Typical wastewater treatment plant (WWTP) installations include UV monitors for process gas monitoring of the ozone concentration, with periodic checks using a wet test meter and an iodometric procedure such as described by Birdsall *et al.* (1952).

Aqueous-Phase Concentration Measurement. The analysis of dissolved ozone residual is difficult because of the transitory nature of ozone in aqueous solution. Ozone tends to disappear quickly in wastewater as a result of reactions with wastewater constituents, autodecomposition, and volatilization. In addition, there are potentially numerous interferences found in wastewater. Dissolved ozone, after being stripped from solution, can be analyzed using gas-phase techniques. Alternatively, ozone residual in wastewater may be measured directly. Reviews of residual oxidant measurement methods have been published by the American Water Works Association Research Foundation (Gordon *et al.*, 1992). Recommended methods include the following (Gordon *et al.*, 1992, and International Ozone Association, 1987a, 1987b, 1987c, 1987d, and 1987e):

- UV absorption,
- Indigo trisulfonate,
- Acid-chrome violet K, and
- Amperometry.

The indigo procedure is a sensitive and precise method for measuring aqueous ozone (Bader and Hoigné, 1981 and 1982, and Bader *et al.*, 1988). It is currently listed as a standard method for ozone residual measurement (APHA, 1995). The indigo method is based on the decoloration of the indigo dye caused by ozone reacting with the double carbon bonds. The resulting decrease in absorbance correlates with the ozone that reacted with the indigo. There are two primary sources of interference with the indigo procedure: chlorine and manganese. A correction is available in the standard method for each interference. Ozone residuals in secondary effluent have been successfully measured using this method (Finch and Smith, 1991).

In the acid-chrome violet K (ACVK) method, the solution containing ozone is added to a solution of the reagent ACVK [1,5-bis-(4-methylphenylamino-2-sodium sulfonate)-9,10-anthraquinone]. The reagent is rapidly decolorized by ozone. The ACVK procedure may be as useful as the indigo procedure, but it has not been adopted as a standard method as yet (Gordon *et al.*, 1992).

Amperometry uses the electrochemical cathodic reduction of ozone to oxygen in solution, according to the following equation (Gordon *et al.*, 1992):

$$O_3 + H_2O + 2e^- \rightarrow O_2 + 2OH^- \qquad E° = 2.42 \text{ V} \qquad (6.2)$$

The concentration of ozone correlates with the current measured between the anode and cathode of the measuring cell. Commercial amperometers use either bare electrodes, which measure dissolved oxidants directly, or ozone-selective membrane electrodes, through which ozone diffuses to the cathode (Gordon *et al.*, 1992). A mixing device is typically required for bare electrode amperometers to prevent fouling of the electrodes and facilitate migration toward the cathode. Membrane amperometers do not experience cathode fouling problems but do exhibit strong temperature dependencies (Gordon *et al.*, 1992, and Stanley and Johnson, 1979). It is likely that in the nutrient-rich environment found in wastewater, the membrane electrode approach would be impractical because of membrane fouling. The bare electrode approach has been used successfully for situations in which the electrodes are mechanically cleaned. However, only total oxidant is measured using this method.

TOXICOLOGICAL PROPERTIES. Occupational Exposure. Ozone is considered a toxic chemical with the health effects shown in Table 6.3. Some of the symptoms reported in humans from exposure to ozone include the following (Bollyky, 1979):

- Irritating odor;
- Burning sensation and watering of the eyes;
- Irritation and drying of the mucous membranes of the nose and throat;
- Listlessness, drowsiness, and tiredness lasting days to weeks;
- Shortness of breath, dry cough, and choking;
- Pain and tightness or burning sensation in chest;
- Frontal headache, upset stomach, and vomiting;
- Light-headedness, dizziness, and fainting;
- Generalized shock and unconsciousness; and
- Pulmonary edema, congestion, hemorrhage (generally within a few hours after exposure to 4 or 5 ppm [v/v] or more), and possible death.

Table 6.3   Health effects of ozone (Bollyky, 1979)

Concentration, ppm by volume	Human response
0.015–0.1	Detectable odor
0.1	Irritation of nose and throat
0.5–1.0	Difficult breathing
1.0–10	Pulmonary edema

The odor threshold of ozone for most individuals is approximately 0.01 ppm (v/v) in air. The maximum allowable exposure for an 8-hour period, as proposed by the U.S. Occupational Safety and Health Administration (OSHA), is 0.10 ppm (ASTM, 1990). Humans experience dryness of the throat, respiratory passages, and eyes on continued exposure to concentrations greater than 0.1 ppm (Stokinger, 1959). None of the symptoms associated with an exposure of less than 1.0 ppm are reported to have lasting effects. However, several weeks may pass before dizziness or tiredness disappears after a severe ozone exposure.

Exposure to levels of 1 ppm for periods longer than 30 minutes will produce headaches in most individuals. Human response to concentrations in the range of 1 to 10 ppm is determined by the duration of exposure, sensitivity of the individual, and tolerance development. Temporary asthmalike symptoms can occur from short-term (minutes) exposures to such levels. Long-term exposures (hours), and short-term exposures to higher concentrations, can cause throat irritations, hemorrhaging, and pulmonary edema. Theoretical human exposure limits for various ozone levels and exposure times are shown in Figure 6.3 (Nebel, 1981).

Ozone is toxic and can be dangerous. As such, ozone demands caution from those who use it. However, several factors mitigate the immediate danger to personnel when ozone is used in a WWTP. First, ozone is not a systemic poison, and the effects of exposure can usually be overcome. Second, ozone has a pungent odor that is readily noticeable at concentrations sufficiently lower than the OSHA threshold for health effects (such as that noticed around photocopiers or laser printers). At higher levels, ozone is "self-

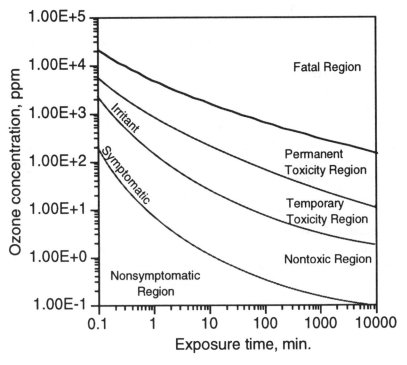

Figure 6.3   Human tolerance for gaseous ozone (Nebel, 1981)

policing" because its pungent odor and immediately noticeable irritation of nasal passages occurs long before any toxic, long-term effects (Diaper, 1972).

Residual Toxicity in the Water Environment. A number of researchers have investigated the toxicity of residual ozone to aquatic life. One of the earliest was Carl Hubbs, who, in 1930, compared changes in water quality following treatment with sodium hypochlorite and ozone (Ward, 1977). In flow-through assays, Hubbs found that residual ozone concentrations of 0.09 mg/L were lethal to fish. In 1975, bioassays were reported on the effects of chlorinated, brominated, and ozonated secondary wastewater effluents on various aquatic species (Arthur *et al.,* 1975). Using special test procedures to induce ozone residuals, these researchers found residual ozone to be approximately as lethal to fathead minnows as residual chlorine.

Ozonated sea water has been reported to have an adverse effect on the commercial American oyster (Maclean *et al.,* 1975). Sea water was filtered to remove large particulate matter, ozonated, and allowed to stand for 2 hours. Initially, no residual ozone was measured, although later studies revealed ozone residuals of 0.18 mg/L after 2 hours and a measurable amount after 4 hours. Use of ozonated sea water decreased fertilization and increased pathogenesis. Whether ozone residual or an oxidized byproduct was responsible was not distinguishable.

Rainbow trout died after exposure to ozonated lake water containing a measurable ozone residual of 0.01 to 0.06 mg/L (Arthur *et al.,* 1975). In a later study, it was reported that long-term exposure to ozone residual levels of 0.016 mg/L or less did not affect the survival, growth, or reproduction of fathead minnows (Ward *et al.,* 1976). However, fingerling trout died within 5 hours when exposed to ozonated effluent containing 0.32 mg/L residual ozone (Ward *et al.,* 1976).

The effects of residual ozone appear to be varied and dependent on many factors. Table 6.4 summarizes the reported effects of residual ozone on a range of aquatic organisms. Although residual ozone has been shown to be toxic, it is, by comparison, no more so than residual chlorine. Also, residual ozone disappears rapidly in the environment. At dosages used in wastewater disinfection, any residual ozone would be short lived and pose few problems to the receiving environment. In fact, because of the enhanced oxygen levels, ozonated effluents may actually be beneficial to the receiving environment once the ozone has dissipated.

Table 6.4   Effects of residual ozone on aquatic organisms (Ward, 1977)

Organism	Residual ozone, mg/L	Summary of effects	Reference
Seven genera of freshwater protozoa	Unknown low concentrations	Death	(Thurberg, 1975)
Rotifer	Unknown low concentrations	Death	(Thurberg, 1975)
Sea urchin and marine worm eggs	Unknown low concentrations	Membrane and cortical changes	(Thurberg, 1975)
Commercial American oyster	Very low or none	Fertilization; developmental and genetic changes	(Dawson et al., 1974)
Rainbow trout	0.01–0.96	Died within 4 hours	(Sengupta et al., 1975)
Barnacle	0.4 and 1.0	Died in several days to a week	(Mangum and McIlhenny, 1975)
Marine phytoplankton	0.1–1.0	Populations decreased in 24–48 hours after exposure to ozone	(Pichet and Hurtubise, 1975)
Crab zoea	0.08 (1 minute)	20% mortality within 24 hours; 30–40% within 48 hours	(Pichet and Hurtubise, 1975)
Crab megalops	0.2	Died within 24 hours	(Pichet and Hurtubise, 1975)
Atlantic silversides	0.08–0.2 (5 minutes)	Died within 30 minutes	(Pichet and Hurtubise, 1975)
Fathead minnow	0.2–0.3	Died within 3 hours	
Fathead minnow	0.00–0.43 ($\bar{x}^a = 0.23$)	Died within 23 hours	
Lake trout	0.18–0.28 ($\bar{x} = 0.23$)	Died within 45 minutes	
Lake trout	0.03–0.06 ($\bar{x} = 0.05$)	Died within 5 hours	
Goldfish	0.222	Died within 72 hours	

[a] $\bar{x}$ = mean concentration.

DISINFECTION

MECHANISMS. Disagreement exists as to whether ozone disinfection occurs by direct ozone oxidation or by reaction with the radical byproducts of ozone decomposition. Hoigné and Bader (1975 and 1978) stated that disinfection was the result of direct ozone oxidation and not the radical mechanism as suggested elsewhere (Dahi, 1976). Some authors have postulated that the disinfection reaction was hydroxyl-radical-mediated, but they have not provided persuasive evidence in support of their hypothesis (Bancroft *et al.*, 1984). It has been reported that pH has little or no effect on ozone disinfection of water, which suggests that the direct ozone reaction predominates (Ross *et al.*, 1976, and Venosa, 1972). Another study reported that *Escherichia coli* (*E. coli*) and *Giardia muris* undergo greater inactivation when ozone residuals persist than when ozone is rapidly decomposed (Finch *et al.*, 1992, and Labatiuk *et al.*, 1994).

In the 1970s, wastewater disinfection requirements were based on the need to reduce fecal coliform to less than 200 colony forming units (CFU) per 100 mL before discharge. The disinfection design criteria were defined by the measurable effluent coliform concentration. Awareness of disinfection processes was heightened by the Safe Drinking Water Act of 1986, in which the need for rational disinfection process design criteria was emphasized. The *CT* (product of residual concentration and time) concept was used as the basis of the engineering requirements for chemical disinfectants.

The *CT* concept is based on work performed in the early 1900s by Harriet Chick and Herbert Watson (Chick, 1908, and Watson, 1908). They thought of the disinfection process as approximately a first-order reaction, with respect to surviving bacteria, if the concentration of disinfectant was held constant throughout the contact time. Their independent approaches have been combined and termed the Chick-Watson model, which is expressed as follows (Haas and Karra, 1984):

$$\log \frac{N}{N_o} = -kC^nT \tag{6.3}$$

The relative effect of concentration on the inactivation of microorganisms is accounted for by n, the dilution factor. Plotting concentration C versus time T on log-log paper for a given inactivation level yields a straight line of slope n (Fair *et al.*, 1947 and 1948). The significance of n can be summarized as follows (Hoff, 1987, and Langlais *et al.*, 1991):

- If $n = 1$, the product of concentration and time (*CT*) remains constant regardless of the concentration of disinfectant;
- If $n > 1$, concentration is the dominant factor for inactivation; and

- If $n < 1$, contact time predominates.

The *CT* values are obtained from experimental data, typically based on 99% inactivation (Hoff, 1987). The plots are then extrapolated for other conditions or inactivation levels, assuming $n = 1$. It is presumed that maintaining a constant *CT* value will achieve the desired inactivation level.

There are several problems inherent in this method:

- Disinfection survival curves typically display shoulder and tailing phenomena, rather than the straight line predicted by the Chick-Watson model;
- The *CT* concept assigns equal importance to concentration and contact time;
- Extrapolating to different conditions may lead to errors; and
- Using the *CT* concept for design can result in costly overdesign of an ozonation system because ozone does not follow the Chick-Watson model and often exhibits an apparent two-stage kinetic curve with a significant tailing effect (Duguet *et al.*, 1989; Finch *et al.*, 1988; Holluta and Unger, 1954; and Katzenelson *et al.*, 1974).

Other models have been proposed that attempt to address some of the shortcomings of the Chick-Watson approach (Haas *et al.*, 1994). An improved rational approach that accounts for disinfectant decay and deviations from the *CT* concept has been reported (Haas *et al.*, 1994, and Haas and Joffe, 1994).

While the *CT* concept and its alternatives provide a rational framework within which to understand chemical disinfection processes, the goals and objectives of wastewater disinfection do not require such a rigorous, analytical approach under current regulations. However, if there is a concern about the efficacy of ozone for inactivating microorganisms other than coliform bacteria, viruses, and protozoan cysts, the reader is directed to the literature available on the topic for drinking water treatment, such as that available from the American Water Works Association Research Foundation in Denver, Colorado.

FACTORS AFFECTING DISINFECTION EFFICACY. The main factors that influence ozone disinfection of wastewater include the following (Bell and Smith, 1982; Caverson *et al.*, 1986; Given and Smith, 1979; and Venosa *et al.*, 1980):

- Ozone dose,
- Contact time,
- Suspended solids,
- pH,
- Chemical oxygen demand (COD),
- Organic carbon,

- Temperature, and
- Contactor design.

Ozone dose and contact time are fundamentally important factors in an effective wastewater disinfection system. Adequate contact time is important to ensure complete mixing of ozone and wastewater (Finch and Smith, 1991). The ozone dose must be sufficient to satisfy the reactions with inorganic and organic species that are faster than the reaction with microorganisms. Once this initial ozone demand has been satisfied, the disinfection reaction with coliform bacteria proceeds rapidly.

Suspended solids increase the ozone demand and shield organisms from the effect of ozone (Venosa *et al.,* 1980). Nevertheless, most secondary effluents can be adequately disinfected without tertiary filtration or other means of polishing. If, however, tertiary polishing processes are anticipated, the ozone demand will decrease and disinfection performance will improve.

Temperature, COD, pH, and organic carbon level affect the rate of ozone decomposition in the wastewater and the total ozone demand. Figure 6.4 illustrates the benefits of removing ozone-demanding substances from the effluent before ozonation. Temperature effects are noteworthy when dealing with ozone in water. Cold temperatures enhance the life of ozone, whereas higher temperatures increase the rate of kill (Labatiuk *et al.,* 1992). The effect of pH on disinfection efficiency is minor (Ross *et al.,* 1976, and Venosa, 1972).

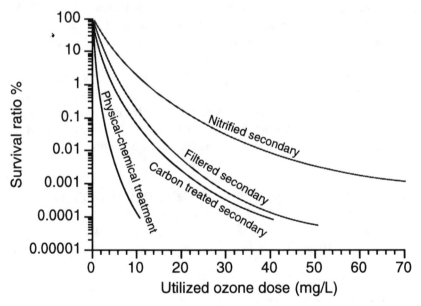

Figure 6.4   **Ozone disinfection of total coliforms in various process effluents (Ghan *et al.,* 1977)**

Contactor characteristics such as poor mass-transfer efficiency, inadequate mixing, and hydraulic short-circuiting adversely affect the performance of disinfection by preventing dissolved ozone and the microorganisms from coming in contact with one another. Proper contactor design will guard against these shortcomings.

Effect on Bacteria. Ozone "can consistently remove up to 99% of the coliform bacteria in a well oxidized secondary effluent at reasonable dosages (10 mg/L or less)" (Ghan *et al.,* 1977). Ghan *et al.* also documented the effects of various process effluents on the ozone consumed during disinfection (Figure 6.4). A contact time of 1 minute after an ozone dose of 5.5 to 6 mg/L reportedly produced a secondary effluent with fecal coliform counts of less than 2 CFU/100 mL (Novak *et al.,* 1977). It has also been reported that relatively low doses of ozone (1.75 or 3.5 mg/L) were sufficient for municipal secondary effluent disinfection, provided the ozone contacting time was followed by a 10- to 20-minute holding time (Bollyky and Siegal, 1977, and Finch and Smith, 1991).

With advances in analytical chemistry in measuring ozone residual, it has been possible to detect ozone residual concentrations in wastewater directly (APHA, 1995). Irrespective of contact time and batch of wastewater, it has been reported that detecting a small ozone residual (0.2 mg/L) at the end of the mass-transfer stage of the contactor reduced *E. coli* concentrations by 4 log cycles (Figure 6.5) (Finch and Smith, 1991). Figure 6.5 illustrates *E. coli* inactivation in a 100-L, semibatch, stirred tank reactor as a function of ozone residual.

The evidence from the literature is contradictory regarding the mode of action of ozone. Some studies have indicated that ozone alters proteins and the unsaturated bonds of fatty acids in the cell membrane, adversely affecting permeability (Pryor *et al.,* 1983, and Scott and Lesher, 1963). However, others have provided evidence of ozone affecting deoxyribonucleic acid (DNA) in the cell, causing cell inactivation (Hamelin and Chung, 1974; Hamelin *et al.,* 1977; Ishizaki *et al.,* 1987; and Ohlrogge and Kernan, 1983). Evidently, the mode of action of ozone is complex, and further research is necessary before the mode of action can be clearly defined.

Effect on Viruses. An excellent review of the literature concerning virus removal by various wastewater treatment processes was published in the early 1970s (Pavoni and Tittlebaum, 1972). It reported that removal efficiencies ranged from minor for primary sedimentation processes to 90% for activated-sludge processes. They also cited studies in which viruses survived wastewater treatment processes, including viruses isolated in chlorinated effluents with 0.5 mg/L residual. This observation has been reported by others who have cited cases of enterovirus and infectious hepatitis virus resistance to chlorine disinfection processes (Shuval and Gruener, 1973).

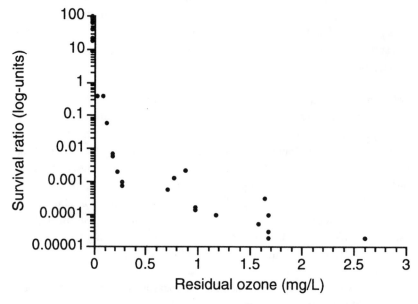

Figure 6.5   *E. coli* **survival as a function of ozone residual in a nitrified secondary effluent (Finch and Smith, 1991)**

Many ozone disinfection studies have been performed using poliovirus as the test organism (Coin *et al.,* 1964; Katzenelson *et al.,* 1974; and Sproul *et al.,* 1979). A group of French researchers found that an approach of 0.4 mg/L ozone residual for 4 minutes could achieve greater than 99.9% reductions (Coin *et al.,* 1964 and 1967). Poliovirus has been reported to be one of the most chlorine-resistant enteroviruses (Feachem *et al.,* 1983). Evidence from the literature suggests that polioviruses are also more resistant to ozone than other enteroviruses (Hoff, 1986, and Safe Drinking Water Committee, 1980). The f2 and MS2 coliphages have been reported to be more resistant to chlorine than other coliphages (Kott *et al.,* 1978, and Tobin, 1987), which is one reason MS2 coliphage has been recommended as a surrogate for enteric viruses in drinking water systems (Malcolm Pirnie, Inc., and HDR Engineering, Inc., 1991).

A study of poliovirus inactivation in ozone-demand free water demonstrated the occurrence of a two-stage inactivation: 99.5% inactivation after 8 seconds and total inactivation after 1 to 5 minutes (Katzenelson *et al.,* 1974). Based on this study, it was proposed that viruses tend to form clumps that are resistant to ozone attacks. Another study found two distinct rates of inactivation with a pivotal point, which was thought to explain the two-stage theory (Majumdar and Sproul, 1974). However, it has been argued by others that these studies were not valid for waters of different qualities and unrepresentative of general applications (Kinman, 1975). In another study that compared poliovirus type 3 and MS2 coliphage, it was found that MS2 was highly sensitive to ozone and was inactivated by greater than 4 log cycles in less than 20 seconds with just a few micrograms of ozone per litre (Finch and Fairbairn, 1991). Poliovirus type 3 was significantly more difficult to inactivate, suggesting that MS2 phage was a poor surrogate for enteric viruses when using ozone (Figure 6.6).

Not many ozone studies have been completed using coliphages (Evison, 1978; Sproul *et al.,* 1979; and Wolfe *et al.,* 1989a and 1989b). In all studies to date, it has been reported that coliphages 185, f2, and MS2 were significantly more sensitive to ozone than were enteric viruses, with 6 to 7 log-unit inactivation occurring quickly and in the presence of little or no measured ozone residual at the end of the contact time. This observation casts doubt on the veracity of coliphages as indicators of ozone disinfection performance for enteric viruses.

Effect on Protozoa. Ozone is a potent disinfectant of the cysts of *Giardia* and *Cryptosporidium* (Finch *et al.,* 1993a and 1993b; Peeters *et al.,* 1989; and Wickramanayake *et al.,* 1984a). While UV radiation is effective against coliforms and other indicator bacteria, UV is not effective against the environ-

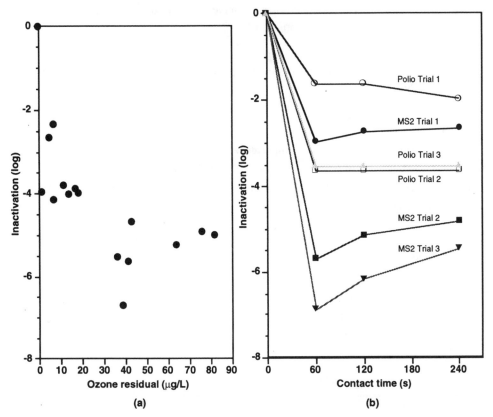

Figure 6.6   Relative sensitivity of MS2 coliphage and poliovirus type 3 (a = MS2 and ozone; b = MS2 and poliovirus) (Finch and Fairbairn, 1991)

mentally robust cysts of intestinal parasites (Lorenzo-Lorenzo *et al.*, 1993, Rice and Hoff, 1981). Therefore, where wastewater-transmitted parasites are a concern, ozone is a disinfectant that should be seriously considered.

Inactivation of *Giardia* cysts with chemical disinfectants has been comprehensively reviewed in three articles (Hoff, 1986; Jakubowski, 1990; and Jarroll, 1988). Until the late 1980s, only one well-documented study on ozone inactivation of *Giardia* had been performed using laboratory water (Wickramanayake, 1984; Wickramanayake and Sproul, 1988; and Wickramanayake *et al.*, 1984a, 1984b, and 1985). This study was the source of the ozone inactivation criteria published in the *Federal Register* (U.S. EPA, 1989) as part of the Surface Water Treatment Rule component of the Safe Drinking Water Act. Since that time, others have suggested that *G. muris* and *G. lamblia* have about the same sensitivity to ozone and that the ozonation requirements may be higher to achieve 99 and 99.9% inactivation of the cysts (Finch *et al.*, 1993b; Metropolitan Water, 1991; Wallis *et al.*, 1990; and Wolfe *et al.*, 1989a and 1989b).

There have been few studies of ozone inactivation of *Cryptosporidium* oocysts to date (Finch *et al.*, 1993a; Korich *et al.*, 1990; Langlais *et al.*, 1990; and Peeters *et al.*, 1989). The reported efforts to achieve greater than or equal to 99% inactivation of oocysts are summarized in Table 6.5.

Table 6.5   **Summary of reported ozonation requirements for inactivation of *Cryptosporidium* sp. oocysts compared with published requirements for *Giardia lamblia***

Species	Ozone residual, mg/L	Contact time, minutes	Temperature, °C	Apparent CT for ≥99% inactivation, mg · min/L	Reference
C. parvum	0.77	6	(Not reported)	4.6	(Peeters *et al.*, 1989)
	0.51	8		4	
C. parvum	1.0	5 and 10	25	5–10	(Korich *et al.*, 1990)
C. baileyi	0.6 and 0.8	4	25	2.4–3.2	(Langlais *et al.*, 1990)
C. parvum	0.17 and 1.9	5, 10, and 15	22	3.5	(Finch *et al.*, 1993a)
G. lamblia	0.11–0.48	0.94–5	5	0.53	(Wickramanayake *et al.*, 1984a)
	0.03–0.15	1.06–5.5	25	0.17	

The mode of action of ozone on cysts is poorly understood but appears to be a surface phenomenon whereby the cyst wall becomes more permeable and uninfective (Finch *et al.,* 1994).

Figure 6.7 illustrates the relative differences in ozonation required to kill *Giardia* and *Cryptosporidium* compared to MS2 coliphage, *E. coli*, and heterotrophic plate count (HPC) bacteria under identical water quality and ozonation conditions in laboratory water. No studies were found on the inactivation of parasites in ozonated wastewater.

DESIGN

FEASIBILITY. To assess the suitability of ozone for a particular location, an evaluation of various disinfectants should be conducted under the expected conditions. Important steps in the evaluation will include

- Conducting a thorough site evaluation and wastewater characterization,
- Reviewing the types of disinfection systems available and their suitability for the site,
- Conducting bench-scale tests to determine the approximate dose of each disinfectant required for the expected wastewater conditions and desired effluent quality,
- Preparing preliminary designs and economic comparisons of the disinfection system based on the results of the bench-scale tests,

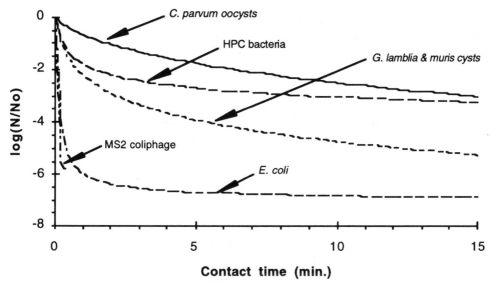

Figure 6.7   Relative susceptibility of MS2, *E. coli*, *Giardia*, *Cryptosporidium*, and heterotrophic plate count bacteria to ozone

- Choosing one or more alternatives for further study, and
- Conducting pilot-scale tests to establish the design and operational parameters.

Bench-Scale Tests. Bench-scale testing can provide information on the following (Schmidtke and Smith, 1983):

- Process feasibility,
- Basic design relationships, and
- Range of disinfectant performance.

The steps involved in bench-scale testing include

- Determining the objectives of the ozonation process;
- Obtaining a sample of the process water to be ozonated (if it is a new WWTP, find a source that has characteristics similar to those expected for the plant);
- By using a simple glass batch reactor, determining the ozone demand of the wastewater over the range of conditions expected (Finch and Smith, 1989; Hoigné, 1994; and Hoigné and Bader, 1994);
- Using a range of ozone doses to determine the ozone dose requirements for achieving the disinfection goals;
- Using factors such as transfer efficiency, maximum ozone dose, and a generator efficiency factor to estimate the size of ozone generator needed for the system; and
- Conducting a preliminary economic analysis of the resulting design.

The water industry has proposed a standardized disinfection testing protocol to help researchers and practitioners approach these complex studies in a consistent manner (Haas *et al.*, 1993).

Pilot-Scale Tests. If bench-scale testing has established that ozone is feasible, then pilot-scale testing should be conducted to refine the parameters required for designing the system. Kinetic parameters established from bench-scale testing can predict the performance of a pilot-scale system, provided the particular wastewater of interest is used (Haas *et al.*, 1994). These parameters include applied ozone dose, transferred ozone dose, transfer efficiency, and initial ozone demand. The variables to be evaluated include gas-flow rate and composition, operating pressure and pressure drop across the contactor, wastewater contact time, desired degree of disinfection, and reaction kinetics.

Design criteria obtained from one gas contactor type are usually not applicable to other types. However, it has been reported that scale-up from a laboratory-scale stirred tank contactor to a pilot-scale diffused bubble column contactor may be successful when using mass-transfer coefficients (Stankovic, 1991). Reactor engineering principles normally base the scale-up of contactors on mass-transfer coefficients; however, because of ozone reaction and decomposition kinetics, this type of scale-up cannot be strictly applied to ozonation systems. It is recommended that the flow regimen of the ozone contactor used in the pilot-scale tests replicate that of the proposed full-scale contactor to achieve results that will be representative of the full-scale design (Stover *et al.,* 1986). The physical arrangement and mixing regimen of the contactor affect the disinfection efficiency. Therefore, it is important to be aware of the geometric, hydrodynamic, and chemical reaction scale-up issues for the system being tested—hence the need to use the same type of contactor during pilot testing.

PROCESS COMPONENTS. Figure 6.8 presents a simplified schematic of an ozonation system. The system consists of four main components: feed gas preparation, ozone generation, ozone contacting, and offgas destruction. Each process step will also require auxiliary equipment and process-control systems. A checklist of the issues to be addressed when designing an ozonation system can be found in *Ozone in Water Treatment: Application and Engineering* (Langlais *et al.*, 1991).

FEED GAS SELECTION AND PREPARATION. Source Gas. Ozone is generated by discharge in an oxygen-containing gas. Process gases that have been used include prepared ambient air, liquid oxygen, on-site gaseous oxygen, or air supplied by means of oxygen-enrichment membranes. In a high-purity oxygen system, the offgas from the contactors may be recycled directly to the aeration basins or treated for reuse in the generators.

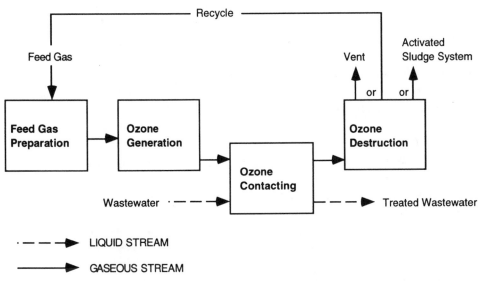

Figure 6.8   **Flow diagram for ozone wastewater disinfection**

The concentration of ozone in the process gas increases twofold to threefold in transferring from air to high-purity oxygen. Therefore, the size of the equipment needed for oxygen-fed ozonators is smaller. Power consumption efficiency is better in oxygen-fed systems than in air systems. An additional benefit of the use of oxygen-fed ozonators is that the capital cost of the contactors may be significantly less when using pure-oxygen systems for ozone generation. Therefore, the economic analysis of an ozone disinfection system should be based on the whole system rather than on any individual components.

When high-purity oxygen is required elsewhere in the WWTP, as in a pure-oxygen activated-sludge process, the economics of using pure oxygen are further improved.

Gas Treatment Systems. Regardless of whether air or high-purity oxygen is used as the feed gas, it must be nominally free of oil, dust, and moisture. The efficiency of ozone generation is directly related to the efficiency with which these parameters are controlled. Oil introduced with the gas stream will cover surfaces such as the desiccant dryers, the piping, and the dielectrics in the generators, reducing efficiency and leading to premature failures. Dust introduced with the gas stream will collect in the ozonator tubes and reduce ozone-production efficiency. Moisture introduced with the gas stream will require an increase in power consumption to maintain the same level of ozone production, and it can reduce the dielectric life. Another key concern is nitrogen, which may be converted to nitric acid and damage the internal parts of the generator in the presence of moisture. If nitrogen is a problem, it may be removed through the use of molecular sieves or other devices.

The typical components of a gas treatment stream are illustrated in Figure 6.9. Treatment of the gas consists of particle removal and some combination of compression, cooling, refrigerant drying, and desiccant drying for moisture removal. Particle removal in these systems may be by filtration or, at larger installations, by electrostatic precipitation. Shell-and-tube heat exchangers may be used as aftercoolers to compensate for the gas temperature increase from gas pressurization. Treated contactor offgas may be returned to the pressurization step, thereby increasing the percentage of oxygen in the feed stream. However, the nitrogen concentration in the offgas must be monitored closely and may prohibit recycling. In addition, a higher desiccant capacity may be needed if the offgas is recirculated.

Compression, cooling, and refrigerant drying may all be used to lower the moisture content of the feed gas, but desiccant dryers are the primary means of moisture removal. Refrigerant dryers can remove moisture using a small amount of energy and will reduce the size of the desiccant dryer needed. However, they require the availability of skilled maintenance personnel and are prone to icing (Stover *et al.*, 1986). Design of the moisture-removal system is based on dew-point temperature, which is the temperature at which the air is

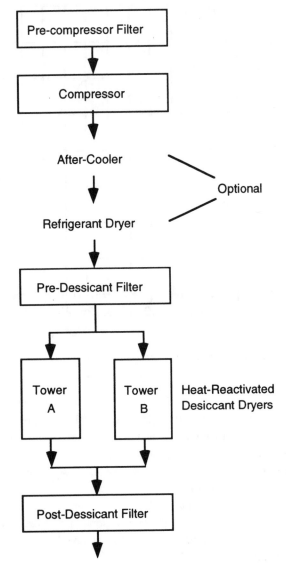

Figure 6.9   Components of low-pressure air feed gas treatment schematic (Stover *et al.*, 1986)

saturated with water for a given pressure. The moisture content of the feed gas increases as the dew point increases. Figure 6.10 illustrates the effect of feed gas dew point on ozone production. The required dew point of the feed gas will depend on the generator design. Recommended practice is that the dew point not exceed a maximum of –60°C (–76°F) (Langlais *et al.*, 1991). Moisture removal is typically not required for pure-oxygen systems because water vapor is not present.

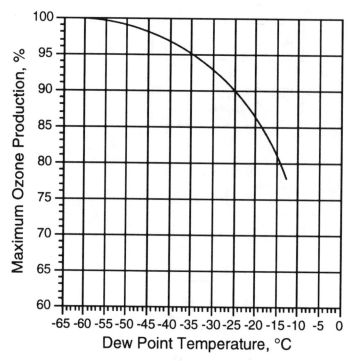

Figure 6.10 **Ozone production versus air dew point (Rakness _et al._, 1988)**

Air treatment systems are often classified by their operating pressures: ambient, low, medium, or high (Langlais _et al._, 1991). A high-pressure system typically requires lower capital cost and maintenance requirements for its desiccant dryers than a low-pressure system, but it requires higher maintenance costs for its air compressors (Stover _et al._, 1986). These advantages and disadvantages must be weighed for each design situation.

The gas stream preparation train should be designed so that sufficient process monitoring and control devices are incorporated. Gas temperature, pressure, flow rate, and oxygen-production rate should be the primary monitoring and control parameters.

Critical elements of the gas-preparation train should be installed in duplicate. This redundancy will allow for maintenance without impairing operation. An example would be the installation of two desiccant drying towers. While one is regenerating, the other may be on line, and vice versa.

OZONE GENERATION. Theory. Ozone is generated when a high-voltage alternating current (6 to 20 kV) is sent across a dielectric discharge gap that contains an oxygen-bearing gas. A simplified schematic of an ozone generator is presented in Figure 6.11. The amount of ozone generated depends on the following (Stover _et al._, 1986):

- Physical characteristics of the generator, including current frequency, voltage, number of dielectrics, dielectric constant, and gap size;
- Power supply to the generator;
- Moisture and dust content of the feed gas;
- Flow rate and oxygen content of the feed gas; and
- Temperature of the gas.

Types of Generators. The three main types of ozone generators are the Otto plate, Lowther plate, and tube types. The Otto plate, although not prominent today, is the oldest type of generator. It generates ozone from dried and cooled air ($-51°C$ [$-60°F$]), at atmospheric pressure. Voltage requirements for this generator range between 7.5 and 20 kV at a frequency of 50 to 500 Hz. Water is used to cool this type of generator.

The Lowther plate generator is air cooled and can generate ozone from both air and oxygen. Feed gas pressures range between 7 and 83 kPa (1 and 12 psig), and peak voltage ranges between 8 and 10 kV at 2 000 Hz. The Lowther plate is not in common use because of inadequate cooling by the air system and maintenance problems (Langlais *et al.,* 1991).

Nearly all current installations use a version of the tube-type generator because of its proven efficiency and reliability. The tube generator is capable of generating ozone from both air and oxygen, which is dried and cooled to between -40 and $-60°C$ (-40 and $-76°F$). Tube generators may be further

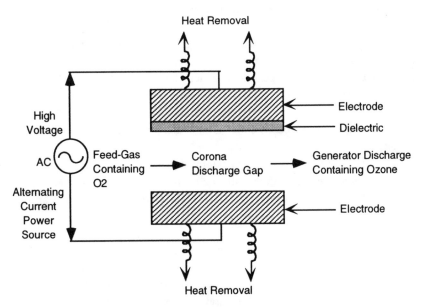

Figure 6.11   **Cross-sectional view of principal elements of a corona discharge ozone generator (Stover *et al.,* 1986)**

classified by the physical arrangement of the dielectrics, the dielectric material, the control mechanism, the cooling mechanism, and the operating frequency. The characteristics of three of the most commonly used arrangements are summarized in Table 6.6.

The frequency ranges used for ozone generators are

- Low, or line, frequency—50 or 60 Hz;
- Medium frequency—60 to 1 000 Hz; and
- High frequency—greater than 1 000 Hz.

Reducing the discharge gap or increasing the frequency will reduce the voltage required and increase the ozone concentration, but these parameters must be balanced with potential losses from dielectric hysteresis, conduction, and dissipation (Langlais *et al.*, 1991). In addition, while decreasing the discharge gap serves to increase ozone production, manufacturing and installation tolerances must be more closely controlled (Langlais *et al.*, 1991).

Ozone concentrations range from 1% (w/w) from air, to 2% (w/w) from oxygen for line-frequency generators, to up to 10% (w/w) from high-frequency oxygen generators. Therefore, gas-flow rates will be lower with higher frequency generators using oxygen as the feed gas. This will have a significant impact on contacting because of reduced gas flows at higher concentrations. Bubble diffuser systems rely on gas-flow rates for mixing in the chambers. Therefore, bubble diffuser contactors are unsuitable when using higher ozone concentrations. Alternative mass-transfer devices such as eductors or injectors can be used.

Optimization of ozone production requires close attention to numerous factors, including gas temperature, ozone concentration, voltage, cooling water flow rate and temperature, generator capacitance, and electrical protection of the dielectric tubes. Generator selection must therefore include consideration of the available technical expertise for operation and maintenance, capital and operating costs, and the ozone concentration and mixing requirements of the contactor.

A number of ozonation process equipment manufacturers have several years experience in successfully using ozone for water and wastewater applications. The reader can consult the International Ozone Association, Stamford, Connecticut, for more information.

Cooling. Ozone generation is inefficient using current corona discharge technology, with a large portion of the energy consumed being converted to heat (Langlais *et al.*, 1991). In addition, the rate of ozone decomposition in the gas phase increases with increasing temperature, as illustrated in Figure 6.12. Consequently, cooling of the ozone generator is an important consideration. Cooling systems may include the following (Langlais *et al.*, 1991):

Table 6.6   Characteristics of tube-type generators

Tube arrangement	Cooling mechanism	Control mechanism	Frequency[a]	Operating gas pressure	Peak voltage[b]	Ozone concentration produced	Comments
Horizontal	Water	Voltage	Low	48–96 kPa	15–27 kV	~1% from air; ~2% from oxygen	Simplest type of water-cooled tube generator; ozone production efficiency is less sensitive to changes in discharge gap than a medium-frequency generator
Horizontal	Water	Voltage	Medium	Depends on discharge gap; generally higher than for low-frequency generator	13–20 kV	Higher than for low-frequency generator	More cooling is required than for a low-frequency generator; ozone produced per energy unit increases as discharge gap is reduced
Vertical	Dual	Frequency	High	Up to 140 kPa	10 kV	Up to 10% from high-purity oxygen	Voltage is kept constant by varying the frequency between 60 and 2 000 Hz

[a] Low frequency = 50 or 60 Hz; medium frequency = 60 to 1 000 Hz; and high frequency = > 1 000 Hz.
[b] Peak voltages vary between frequencies and manufacturers.

- A closed circuit using a plate heat exchanger and recirculation pumps,
- A closed circuit with a chiller, or
- A chiller.

Water cooling is typically used, though air-cooled systems are available. The cooling system, regardless of whether it uses air or water, must be closely monitored. This monitoring requires the permanent installation of temperature sensors and recorders at both the influent and effluent ports of the ozonator. Cooling water should be treated in accordance with proper engineering practice to avoid any scaling, corrosion, or precipitation in the generator or piping. Effluent wastewater is too impure to be used for cooling.

Gas-flow rate is an important consideration in ozone production and temperature control. If the gas-flow rate is too low, the conversion step will result in excessive temperatures. If the gas-flow rate is too high, the oxygen-to-ozone conversion efficiency will deteriorate. It is therefore imperative to maintain an optimum gas-flow rate as recommended by the manufacturer of the generator.

CONTACTING. Objectives. The introduction of ozone-bearing gas to the wastewater is a critical step in the disinfection process. The main purpose of the gas–liquid contactor is to transfer ozone from the gas phase to the liquid

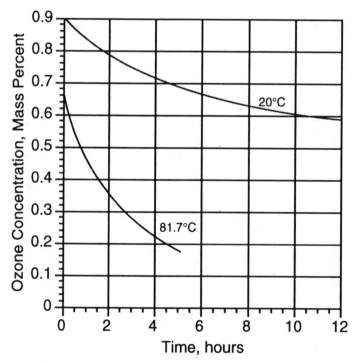

Figure 6.12   Decomposition of ozone in air

phase for reaction with the wastewater constituents (Stover *et al.*, 1986). Three steps are involved: mass transfer of ozone; mixing of the dissolved ozone with the wastewater to ensure that all of the microorganisms are exposed; and a final, low-turbulence stage during which the ozone reacts with the wastewater constituents. Inherent in the design of ozone contactors is the need to prevent hydraulic short-circuiting and minimize areas of under- and overozonation. In the past, attempts have been made to meet all of these objectives in a single stage. However, innovative engineering designs have evolved in which the mass-transfer and mixing stages are separated from the contacting stage. By optimizing the contactor for mass transfer and mixing in the same stage, consistent disinfection performance is possible in the subsequent reactive flow segment (Finch and Smith, 1991, and Lev and Regli, 1992a and 1992b).

GAS MASS TRANSFER. Gas–liquid mass transfer has a well-documented theoretical basis in the literature (Danckwerts, 1970). The reader is directed to other sources for a complete description of gas–liquid reactions (Grasso *et al.*, 1990; Mariñas *et al.*, 1991; Rakness *et al.*, 1988; Roustan *et al.*, 1987; Stankovic, 1991; Zhou and Smith, 1994; and Zhou *et al.*, 1994).

Three mass-transfer situations arise in practice:

- Mass transfer is limited by the rate of the reaction in the bulk solution;
- Mass transfer is limited by the rate of mass transfer at the liquid–gas film in solution; or
- Mass transfer is enhanced by the rapid reaction of ozone and wastewater.

In most cases, ozone disinfection systems are mass-transfer limited rather than reaction-rate limited. Consequently, sizing of contactors is usually contingent on the rate at which ozone is transferred to the bulk liquid. One report (Robson, 1982) has shown that the reaction with constituents found in wastewater increases mass-transfer efficiency. However, another report (Chang and Singer, 1991) has shown that the rate of mass transfer in most ozonation applications is probably not enhanced by the liquid-phase oxidation reaction and can be considered equivalent to the rate of a purely physical mass transfer.

The major impetus for mass transfer is a concentration gradient described by

$$\frac{d[O_3]}{dt} = K_L a([O_3]_s - [O_3])$$
(6.4)

Where

$[O_3]$ = ozone concentration at time t, seconds;
$K_L a$ = volumetric mass-transfer coefficient, 1/s; and
$[O_3]_s$ = ozone saturation concentration, mg/L.

The gradient is maintained by constantly supplying ozone to the high end of the gradient and removing it at the low concentration end. Good bulk transfer of ozone into the liquid stream promotes good absorption. Primarily, this depends on the intimate contact between ozone and wastewater under conditions that promote dissolution.

In practice, ozone mass-transfer efficiency is governed by several factors, including

- Concentration of ozone in the gas phase,
- Gas-flow rate,
- Mixing conditions during the mass-transfer stage,
- Bubble size,
- Shear at the gas–liquid interface,
- Geometry of the contactor and arrangement of the gas spargers, and
- Wastewater quality.

DISINFECTION EFFICIENCY. Many factors that arise in practice can adversely affect the performance of wastewater disinfection when using ozone, including the following (Stover *et al.,* 1986):

- Ozone-transfer efficiency,
- Short-circuiting in the ozone contactor,
- Mixing, and
- Contact time.

The major problem of disinfection processes is hydraulic short-circuiting, whereby the actual retention time is significantly less than the theoretical retention time. It has been shown that a short-circuiting of only 2% may result in effluent coliforms of 2 200 CFU/100 mL, as opposed to 200 CFU/100 mL with no short-circuiting (Stover *et al.,* 1986).

Short-circuiting may occur for a variety of reasons and is largely dependent on the contactor type and configuration. Single-stage or two-stage bubble diffuser contactors are sensitive to short-circuiting because of the dependency of the gas-flow rate on the liquid-flow rate and the use of gas flow for mixing. During high gas flows, channeling of the bubbles will occur. High wastewater-flow conditions may also cause short-circuiting. At the other extreme, low gas-flow conditions may also cause short-circuiting because mixing is poor. Optimizing gas- and wastewater-flow rates using appropriate computer-aided process control measures will minimize short-circuiting and increase the number of stages to at least three or more with good baffling (Rakness *et al.,* 1993).

Although mixing the contents of the reactor is necessary, turbulent mixing after the ozone has been transferred to solution will volatilize the ozone and stop the reactions (Farooq, 1976, and Finch and Smith, 1991). Because of the rapid reaction kinetics of ozone, contact time is not as critical a parameter as it is in chlorination systems. Typical contact times range between 2 and 10 minutes. The optimum contact time should be established through pilot testing (Stover *et al.,* 1986).

Types of Contactors. The types of contactors commonly used for gas–liquid contacting are summarized in Table 6.7. Illustrations of three common ozone contactors are provided in Figures 6.13 to 6.15. The choice of contactor will be determined by several factors, including

- Degree of diffuser clogging that can be anticipated,
- Concentration of ozone in the feed gas,
- Required contact time,
- Need for offgas collection and reuse,
- Whether the reactions are mass transfer or rate limited, and
- Constraints of the site.

Much research has been done in the evaluation of contactors. An early study compared a cocurrent diffused bubble contactor, positive pressure injector, and batch reactor. Scaccia and Rosen (1978) concluded that all types were suitable provided they were properly designed. Another group evaluated a submerged turbine, a mechanical surface mixer, and two bubble diffusers (Stover *et al.,* 1987). The first bubble diffuser operated in countercurrent flow, whereas the second bubble diffuser consisted of a countercurrent first stage followed by a cocurrent second stage. They found that all four of the contactors could perform effectively and that consideration of factors such as initial ozone demand, water quality, and gas-flow rate will determine which contactor performs best.

A similar study reported that the counter-current bubble diffuser was the most efficient and efficacious (Venosa *et al.,* 1978). Indeed, the counter-current bubble contactor has found the widest application in wastewater

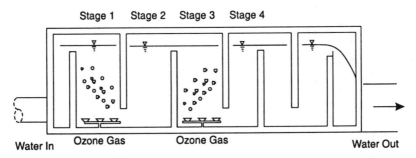

Figure 6.13 **Schematic of a four-stage bubble diffuser ozone contact basin (for low gas-phase ozone concentrations) (Langlais *et al.,* 1991)**

Table 6.7 Summary of gas–liquid contacting systems (Langlais et al., 1991)

Type	General characteristics	Advantages	Disadvantages
Conventional fine-bubble diffusion	Most widely used type of contactor Various configurations, baffle arrangements, and flow patterns are available; cocurrent and countercurrent flow are two common examples; mixing is achieved solely by the movement of the gas bubbles	No further addition of energy is required after initial gas compression No moving parts; therefore, minimal maintenance is required Can achieve ozone-transfer efficiencies greater than 90% Scale-up is less difficult because these systems are widely used and understood	Relatively deep contact basins are required for effective ozone transfer Potential clogging of diffuser pores Vertical channeling of bubbles can occur
Turbine mixers: positive gas pressure and negative gas pressure	Also quite commonly used May draw the gas into the chamber as well as promote mixing	High turbulence and heat of turbine creates small bubbles—increases transfer efficiency Less depth is required than for bubble diffusers to achieve effective ozone transfer May be used to draw offgases from later contact chambers into the initial chamber, where ozone demand is higher Can achieve mixing and ozone contacting in one chamber Diffuser clogging problem is alleviated Scale-up is less difficult because these systems are widely used and understood	Energy is required to run the turbine Must maintain a constant gas flow rate, regardless of the water flow rate; cost per volume increases and transfer efficiency decreases as water rate decreases Submerged systems may be difficult to maintain

Table 6.7   Summary of gas–liquid contacting systems (Langlais *et al.*, 1991) (continued)

Type	General characteristics	Advantages	Disadvantages
Injectors and static mixers: positive gas pressure and negative gas pressure	Ozonized gas is injected into the liquid stream, either under pressure or via a partial vacuum (created by flowing the liquid quickly past a small orifice, drawing the gas through the orifice) Downstream static mixing may be required following injection	No moving parts; therefore, minimal maintenance is required Can achieve good mixing and mass transfer of ozone Contractor size may be smaller than required for bubble diffuser Ozone utilization rate may be higher than for conventional countercurrent contactor	Head loss must be overcome by pumping Short contact times—may need to follow the injection or mixing point with a reaction chamber Ability to turn down the system for lower flow rates is limited
Packed columns	Use for wastewater treatment has been limited Two approaches: add ozone prior to column and use column as a reactor, or add the ozone within the column, using the column as both a reactor and a mixer	No moving parts; therefore, minimal maintenance is required Scale-up for hydraulics and head loss are well understood The packing creates a flow regime that is more representative of plug flow; this is an advantage because plug-flow reactors require less contact time than CSTRs to achieve the same reaction efficiency Lower inlet gas pressures are required	Scale buildup may occur, causing increased head losses Experience regarding scale-up for ozone transfer and treatment efficiencies is limited
Spray chambers	Wastewater to be treated is sprayed into ozone-rich air		Contact times are short; therefore, spray chambers are only suitable for rapid reactions

Table 6.7   Summary of gas–liquid contacting systems (Langlais *et al.*, 1991) (continued)

Type	General characteristics	Advantages	Disadvantages
Deep U-tube	Consists of two vertical concentric tubes; water and ozone travel downward through the inner tube, and then flow upward through the outer tube	The external pressure on the gas bubbles increases as the water and gas flow downward in the inner tube; this enhances the ozone mass transfer The mixing regime is turbulent, enhancing ozone transfer; does not clog Operates as a plug-flow reactor The contactor is quite deep (20 m [66 ft]); thus, surface area requirements are less than for other contactors	Experience is limited; needs to have auxiliary pump to overcome head loss problems Ability to turn down the system for lower flow rates is limited because high velocities are required to deliver the gas to the bottom of the tube Proprietary design More challenges may arise with installing the contactor because of its depth (for example, excavation) Maintenance may be more difficult
Submerged static radial turbine	Still in developmental stages A portion of the wastewater is pressurized and mixed with gas in the turbine, then injected into the contactor through nozzles; fine bubbles are created because of high shear Operates in countercurrent mode	Does not contain submerged moving parts, so maintenance may be easier than for conventional turbines Useful when space limitations do not allow for use of a bubble contactor Good ozone-transfer efficiencies are possible	Experience is limited because this is a new system

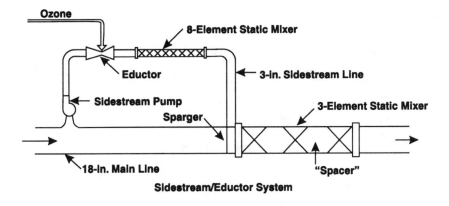

Sidestream/Eductor System

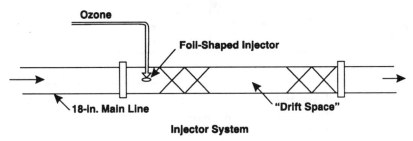

Injector System

Figure 6.14   Schematic of a gas induction system with in-line static mixer (for high gas-phase ozone concentrations) (in. × 25.44 = mm) (James M. Montgomery, 1989)

disinfection. The countercurrent configuration provides superior ozone absorption efficiency and delivers a higher concentration of ozone to subsequent contact chambers than cocurrent flow conditions (Zhou *et al.,* 1994).

Contactor Design Considerations.   Selection of the appropriate contactor must include the following considerations: mixing conditions, water quality, ozone-transfer efficiency, ozone decay, and gas concentrations and flow rates. The practical considerations and guidelines in current use are briefly described below for some of the common types of contactors. Detailed information can be obtained from the International Ozone Association and other sources (Langlais *et al.,* 1991).

FINE-BUBBLE DIFFUSERS.   Current practice for fine-bubble diffusers (Figure 6.13) is to use two or more countercurrent stages in series. After the contactor has begun operation, tracer studies should be conducted to establish the actual contact time versus the theoretical hydraulic retention time. Baffles are commonly used to decrease short-circuiting.

Water depths above the diffusers range from 4.88 to 6.10 m. Deeper contactors are used at higher elevations to account for reductions in

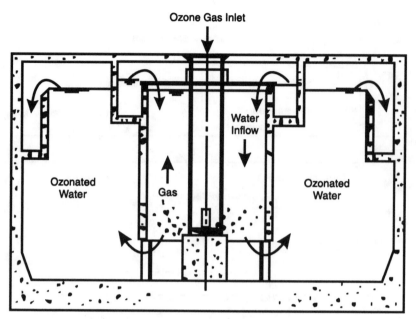

Figure 6.15   Schematic of a submerged turbine mixer ozone contactor (for low through high gas-phase ozone concentrations) (Langlais *et al.,* 1991)

ozone-transfer efficiency with increased elevation. There are a variety of diffusers (such as glass, fritted, and ceramic) available for contactors. They should be evaluated based on their capacity to promote mass transfer and reduce the possibility of ozone decomposition. Fine bubbles will provide a larger surface area for disinfection reactions. In wastewater applications, the potential for clogging should be evaluated, and planned maintenance is necessary.

Two important considerations in the design of fine-bubble diffusers are the gas-to-liquid ratio and gas-flow rate. As the ratio of gas to liquid increases, ozone-transfer efficiency becomes limited. This is thought to be caused by interfering and coalescing of the bubbles. The recommended upper limit for the gas-to-liquid ratio is 0.15:0.20.

The gas-flow rate is important because it directly affects the potential for short-circuiting in the contactor. Within the contactor, the liquid takes the path of least resistance, that is, wherever gas bubble flow is low. Hence, there is a greater potential for short-circuiting at low gas-feed rates. This most commonly occurs when high concentrations of ozone are applied, usually with pure-oxygen systems. Too few diffusers is another cause of low-flow conditions. A minimum gas-flow rate of 2.19 m^3/m^2•h is recommended to ensure adequate mixing.

The following suggestions have been made to accommodate both the gas-flow rate and gas-to-liquid ratio constraints (Langlais *et al.,* 1991):

- To meet the minimum gas-feed rate, it may be necessary to decrease the ozone concentration and maintain a constant gas flow;
- If the gas-to-liquid ratio is too high, the ozone application could be divided into several successive stages in series;
- If neither of these is sufficient, the contactor may be divided into parallel units and the units taken off line, as required for low-flow conditions; and/or
- It may be necessary to use a different type of contactor for the design, such as a turbine or injector system, for which mixing is inherent in the design.

INJECTOR AND STATIC MIXER. For an injector and static mixer (Figure 6.14), the design objective is to ensure that the ozone will remain in solution, provided the pressure remains constant throughout the system. As the ozone/oxygen concentration increases, so does the saturation limit. The designer must therefore evaluate the gas saturation solubility in the water to be treated. Sufficient contactor time, either in-line or in a contactor, must also be provided. This type of contactor is becoming increasingly attractive because of its high mass-transfer efficiencies, good oxidation reactions, and relatively low capital construction costs (Martin *et al.,* 1992). For example, the outfall pipeline can be used as the contactor once the ozone has been injected and mixed.

Sidestream injection is more effective than direct injection for the following reasons: higher transfer efficiency because of increased mixing, higher ozone-utilization rate, and reduced contact time requirement.

TURBINES. The design of a turbine contactor (Figure 6.15) depends on the type of turbine. There have been a number of turbine installations for situations in which the water quality leads to rapid clogging of diffusers. Also, when using high-concentration ozone generators, the ozone must be mixed vigorously to promote adequate mass transfer. Because most turbine systems are patented, the various manufacturers should be consulted.

Process Control. There are several strategies available for process control in wastewater disinfection systems. They all attempt to monitor ozone or a surrogate in the contactor with feed-back loops to the ozone-generation module. An excellent approach has been to monitor the offgas concentration from the contactor (Venosa *et al.,* 1985). This has been shown to adequately track the ozone requirements of the wastewater (Rakness *et al.,* 1984 and 1993). A second approach is to use electrochemical or chemical ozone residual measurement techniques. Either approach is suitable for process control of wastewater disinfection. However, the offgas procedure has fewer operational problems than the direct measurement of ozone residual.

OFFGAS DESTRUCTION. For environmental and safety reasons, ozone remaining in the contactor offgas must be destroyed before the gas can be vented to the atmosphere. The three main methods are thermal, thermal/ catalyst, and catalyst. Each of these techniques is described further below. Some other, less common, techniques include activated carbon adsorption, dilution and venting, and chemical destruction (Langlais *et al.*, 1991).

Thermal ozone destruction involves heating the gas to between 300 and 350°C for up to 5 seconds (Langlais *et al.*, 1991). Heat exchangers are used to recover heat from the discharged gas stream and then preheat the gas stream entering the destruction unit. This saves on the energy required to destroy the gas as well as avoiding the discharge of gas at an elevated temperature. Thermal destruction systems are not recommended for use with high-purity oxygen systems because of the potential fire hazard (Stover *et al.*, 1986).

Catalytic destruction uses the ability of metals, metal oxides, hydroxides, and peroxides to decompose ozone. The actual composition of specific catalysts is proprietary information; therefore, a manufacturer's input is essential when sizing the unit. The offgas is first passed through a demister and then contacted with the catalyst in a chamber. Moisture removal is essential to ensure proper functioning of the catalyst.

Heating of the gas before it enters the catalyst chamber is recommended to prevent moisture condensation on the catalyst (Stover *et al.*, 1986). A lower temperature is required than for thermal destruction acting alone. A typical range for a metal oxide catalyst is 50 to 70°C.

MATERIALS. Because of ozone's strong oxidizing power, all materials of construction downstream of the generator must be resistant to oxidation (that is, made of stainless steel 304 [L] or 316 [L], glass, or polytetrafluoroethylene). Table 6.8 is a listing of materials resistant to corrosion (Damez, 1982). All pipe connections should be welded when possible; otherwise, the connections should be either threaded or flanged.

A check valve should be installed in the gas-feed line ahead of the contactor. This check valve will prevent a major catastrophe if the gas supply should be cut off and the liquid backs up into the gas line. The valve cracking pressure (the pressure at which the valve first opens to allow flow) is an important design consideration. High cracking pressures should be avoided.

PROCESS DESIGN EXAMPLE. The design of an ozone disinfection system can take two approaches (Stover *et al.*, 1986):

- Review the required applied ozone doses and transfer efficiencies at similar existing WWTPs and base the design on these values; or
- Use a rational approach, whereby the design is based on the relationship between coliform removal and transferred or utilized ozone dosage, as established from bench-scale and/or pilot-scale tests.

Table 6.8 Materials resistant to corrosion (from Damez, F. Materials Resistant to Corrosion and Degradation in Contact with Ozone. In *Ozonation Manual for Water and Wastewater Treatment.* W.J. Masachelein (Ed.); Copyright © 1982; reprinted by permission of John Wiley & Sons, Ltd., New York, N.Y.)

Materials	Ozonation, type of exposure			Comments
	Dry air	Moist air	Water	
Metals				
Chromium–nickel–silver	b	b	b	
Brass	b	b	b	
Aluminum	a	d	d	
Aluminum alloys	a	d	d	
Pig iron	a	a	a	Slow corrosion
Galvanized steel	b	c	c	Not shock resistant
Stainless steel	a	a	a	Without chlorine present
Sintered stainless steel	d	d	d	
Plastics				
PVC[a] supple	b	b	b	
PVC rigid	a	a	a	Without pressure
Vinyl ester resin	a	a	a	
Polytetrafluoroethylene (Teflon™)	a	a	a	
Polyamide (nylon-risen)	—	—	a	
Epoxide (Araldite™)	—	d	d	
Synthetic rubbers				
Chlorosulfonated polyethylene (Hypalon™)	a	a	a	With appropriate charge
FPM (viton)	b	c	c	With appropriate charge
Silicone	d	c	c	With appropriate charge
Ethylene propylene	a	a	a	With appropriate charge
Polychloroprene (neoprene)	b	c	c	With appropriate charge
Concrete	a	a	a	
Glass and ceramics	a	a	a	

Note: a = long lasting; b = usable; c = for low ozone concentration; and d = quickly degraded.

[a] PVC = polyvinyl chloride.

The second approach is the recommended method because the wastewater characteristics and contactor behavior can vary greatly between WWTPs. The steps involved in this method and example calculations, as outlined by Stover *et al.* (1986), are discussed below.

A plot of log coliform survival versus utilized ozone dose yields essentially a straight line after the initial demand has been satisfied (Finch and Smith, 1989 and 1990). The general form of the equation of the line is

$$\log(N/N_o) = n \times \log(U/q) \tag{6.5}$$

Where

N	=	effluent coliform concentration, CFU/100 mL;
N_o	=	influent coliform concentration, CFU/100 mL;
n	=	slope of dose-response curve;
U	=	utilized ozone dose (mg/L); and
q	=	x-axis intercept of dose-response curve, which is assumed to equal the initial ozone demand.

A high value for the x-intercept correlates to a high initial ozone demand and will lead to a much higher utilized ozone dosage required for the system. Likewise, the slope of the dose-response curve represents the relationship between utilized ozone and coliform survival. Hence, a steeper slope translates to a lower required dose to achieve a particular level of kill.

The process design conditions that are used for this example are summarized in Table 6.9. Disinfection data have been collected in bench and pilot scales using methods similar to other studies reported in the literature (Finch and Smith, 1989). Because there was no statistically significant difference between the bench- and pilot-scale data, all of the data have been used for the dose-response equation. The bench- and pilot-scale results are tabulated in Table 6.10 and graphically illustrated in Figure 6.16. These data are used to estimate the model parameters for Equation 6.6 below.

Contact time beyond 1 or 2 minutes is not important as a process design parameter for inactivating coliform bacteria because the reaction is fast (Finch and Smith, 1989, and Stover and Jarnis, 1980). Others have observed that 10 to 15 minutes of contact time is sufficient for coliform inactivation based on hydrodynamic considerations rather than chemical-reaction kinetics (Stover *et al.*, 1986, Stover and Jarnis, 1981).

Table 6.9 Ozone disinfection system criteria for example design problem

Average daily wastewater flow	28 ML/d (7.5 mgd)
Peak daily wastewater flow	57 ML/d (15.0 mgd)
Start-up daily average wastewater flow	13 ML/d (3.5 mgd)
Start-up peak daily wastewater flow	28 ML/d (7.5 mgd)
Average effluent BOD[a]/TSS[b]	15/15 mg/L
Maximum daily effluent BOD/TSS	30/30 mg/L
Design required effluent fecal coliform	
Weekly maximum limitation	400 per 100 mL
Geometric mean monthly limitation	200 per 100 mL
Disinfection system influent fecal coliform	
Geometric mean concentration	500 000 per 100 mL
Maximum concentration	2 000 000 per 100 mL

[a] BOD = biochemical oxygen demand.
[b] TSS = total suspended solids.

Table 6.10 Summary of dose-response data collected for ozone disinfection of *E. coli* in a nitrified secondary effluent

Scale	Dose level	Mean transferred ozone, mg/L	Mean utilized ozone, mg/L	Mean *E. coli* survival, log N/N_o	Trials, no.	95% confidence limits for mean log N/N_o Lower	95% confidence limits for mean log N/N_o Upper
Bench	1	9.4	9.4	−4.59	5	−4.71	−4.47
Bench	2	4.9	4.9	−4.39	6	−4.50	−4.28
Bench	3	2.4	2.4	−0.496	6	−0.605	−0.387
Bench	4	1.3	1.3	−0.253	6	−0.362	−0.144
Pilot	1	22	21	−6.38	6	−6.71	−6.06
Pilot	2	12	11	−5.34	6	−5.81	−4.86
Pilot	3	6.0	5.7	−4.70	9	−4.94	−4.46
Pilot	4	3.0	3.0	−1.28	6	−2.22	−0.332
Pilot	5	1.5	1.5	−0.083 9	6	−0.150	−0.018 0

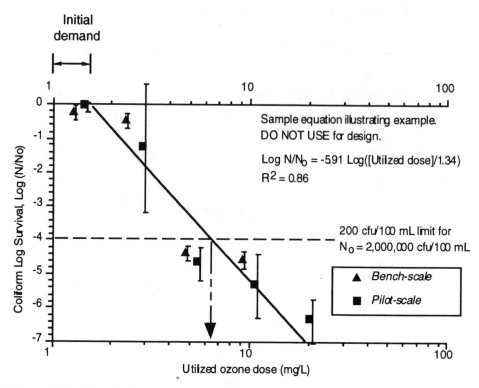

Figure 6.16 Sample dose-response curve for ozone disinfection of *E. coli* in secondary effluent

Step 1: Determine the Initial Ozone Demand and Slope of the Dose-Response Curve. From Figure 6.16, the *x*-intercept is determined to be 1.34 mg/L, and the slope of the curve is –5.91. For various ranges of influent coliform concentration, with a desired effluent coliform concentration of 200 per 100 mL, the utilized ozone dose can then be estimated from Equation 6.5 above:

$$U = q \times 10^{[\log(N/N_o)/n]} \qquad (6.6)$$

Where

q = 1.34 mg/L,
n = –5.91, and
N = 200 coliforms per 100 mL.

Choosing N_o = 2 000 000 CFU/100 mL, the estimated utilized ozone dose U is 6.4 mg/L. This value may also be read directly from Figure 6.16; for a 4-log reduction in coliforms, the utilized dose is 6.4 mg/L, as shown by the dashed lines.

Step 2: Choose the Ozone-Transfer Efficiency and Size the Contactor.
As discussed above, the advantages, disadvantages, and cost of each type of contactor; gas-flow considerations; and objectives of the system must be considered when evaluating the best type of contactor for the application. Ideally, the pilot-scale testing should be conducted using a contactor that models the expected final design. In addition, operational experiences using similar contactors at existing WWTPs should be reviewed for contactor performance. For this example, an air-feed, low-frequency ozone generator has been selected. Therefore, a four-stage, two-basin bubble diffuser has been chosen as a suitable reactor. Mass-transfer efficiency is assumed to be 80%, which is slightly conservative for this type of contactor. (Note: Mass transfer efficiency is a comparison of the feed gas concentration of ozone compared to the offgas concentration of ozone. If the feed gas is 1.2% [w/w] and the offgas is 0.2% [w/w], the transfer efficiency is $[1 - \{0.2/1.2\}] \times 100$, or 83%. The utilized ozone is the difference between the transferred ozone dose [mass of ozone transferred per unit volume] and the observed ozone residual [ozone mass per unit volume]. Typically, the ozone residual is near zero in wastewater disinfection; therefore, the transferred and utilized ozone can be considered the same for all practical purposes.)

The contactor features are as follows:

- Initial ozone demand: 1.34 mg/L.
- Slope of dose-response curve: –5.91.
- Used:applied dose efficiency: 80%.
- Type of contactor: bubble diffuser.
- Number of stages: 4.
- Number of basins: 2.
- Contact basin detention time at average flow: 15 minutes.
- Diffuser depth: 5.5 m.

Step 3: Determine the Design Ozone Production Rate. Using the efficiency of 80%, the required applied dose for each of the expected conditions may be calculated. These calculations are summarized in Table 6.11. Doses ranging from 3 to 8 mg/L are required to meet the effluent coliform objective of 200 per 100 mL. The ozone generator must be capable of producing 332 g/min or 478 kg/d to satisfy all of the conditions.

Step 4: Select the Number of Ozone Generators. The ozone generators must be able to meet both the minimum and maximum required rates. Thus, the ozone production range must fall within a range of 34 to 332 g/min or a turndown ratio of 10. While one large generator can handle the peak output, the turndown ratio greatly influences the operating costs. Prospective manufacturers must be consulted to determine the most economical configurations for the site. In this example, three generators would be suitable for worst-case

Table 6.11 Design example calculations (Langlais *et al.*, 1991)

Condition number	Wastewater flow rate, ML/d	Desired effluent fecal coliform, CFU/100 mL	Influent fecal coliform, CFU/100 mL	Log coliform survival	Utilized dose, mg/L	Applied dose, mg/L, for $E = 80\%$	Applied mass, kg/d	Applied mass, g/min
1	60	200	10 000	−1.70	2.60	3.25	195	136
2	60	200	500 000	−3.40	5.04	6.30	378	263
3	60	200	2 000 000	−4.00	6.38	7.97	478	332
4	30	200	10 000	−1.70	2.60	3.25	98	68
5	30	200	500 000	−3.40	5.04	6.30	189	131
6	30	200	2 000 000	−4.00	6.38	7.97	239	166
7	15	200	10 000	−1.70	2.60	3.25	49	34
8	15	200	500 000	−3.40	5.04	6.30	95	66
9	15	200	2 000 000	−4.00	6.38	7.97	120	83

conditions, leaving a standby generator available for normal operating conditions. When designing the equipment rooms, keep in mind that future changes in wastewater conditions, flow rates, and effluent objectives may lead to a requirement for more ozonation capacity, which can easily be achieved by adding a generator or two.

Step 5: Determine the Size and Number of Air Compressors. Proposed design parameters are

- The maximum ozone concentration in the gas phase is 18 g/m^3 (1.5% w/w);
- A low-pressure, air-fed system is used; and
- The desiccant dryer purge air flow is 20%.

The generator and total system air flow rates must be calculated at average start-up and peak design conditions:

$$\text{Generator air flow rate } (m^3/min)$$
$$= \frac{\text{Applied mass (kg } O_3/d)}{\text{Ozone concentration (kg air/kg } O_3)} \quad (6.7)$$
$$\times \frac{1 \text{ kg air}}{1.21 m^3} \times \frac{1 \text{ d}}{1\,440 \text{ min}}$$

$$\text{Total system air flow rate } (m^3/min)$$
$$= \frac{\text{Generator air flow rate } (m^3/min)}{\text{Percentage of total air flow going to generators}} \quad (6.8)$$

At average start-up conditions,

$$\text{Generator air flow rate } (m^3/min)$$
$$= \frac{95 \text{ kg}O_3/d}{0.015 \text{ kg air/kg } O_3} \times \frac{1 \text{ kg air}}{1.21 m^3} \times \frac{1 \text{ d}}{1\,440 \text{ min}}$$
$$= 3.6 \ m^3/min(128 \text{ scfm})$$

$$\text{Total system air flow rate } (m^3/min)$$
$$= \frac{3.6 \ m^3/min}{0.8} = 4.5 \ m^3/min(160 \text{ scfm})$$

At peak design conditions,

$$\text{Generator air flow rate } (m^3/min)$$

$$= \frac{478 \text{ kg } O_3/d}{0.015 \text{ kg air/kg } O_3} \times \frac{1 \text{ kg air}}{1.21 \, m^3} \times \frac{1 \text{ d}}{1\,440 \text{ min}}$$

$$= 18.3 \ m^3/min \ (647 \text{ scfm})$$

$$\text{Total system air flow rate } (m^3/min)$$

$$= \frac{18.3 \ m^3/min}{0.8} = 22.9 \ m^3/min \ (809 \text{ scfm})$$

Three compressors with capacities of 8.5 m³/min (300 scfm) will provide the peak air flow rate required; however, four compressors are recommended to minimize power requirements at start-up. Stover *et al.* (1986) recommend that two of the compressors be rated at 8.5 m³/min and the others at half of this. In addition, one large and one small compressor should be equipped with a variable-speed drive.

Step 6: Determine the Size and Number of Desiccant Dryers. Proposed design parameters are as follows:

- The operating pressure is 93 kPa (13.5 psig).
- The WWTP is at an elevation of 915 m (3 500 ft) above sea level; atmospheric pressure is 89.1 kPa (12.93 psi); absolute pressure, therefore, is 182 kPa (26.43 psia).
- The inlet air temperature is 30°C (86°F).
- With an aftercooler, the desiccant dryer inlet air temperature is 20°C (68°F).
- The required dew-point temperature, at 1 atm pressure, is –60°C (–76°F).
- The desiccant design cycle time is 16 hours.
- The maximum desiccant loading is 8 kg (18 lb) of water received during one complete drying cycle per pound desiccant (17.6 kg of water per kilogram of desiccant).
- The design gas-flow rate is 25.5 m³/min (900 scfm).

First, the moisture loading to the desiccant dryer must be calculated. From Table 6.12, the moisture content of the 20°C (68°F) desiccant dryer inlet air is 17.8 g/m³ (1.110 lb/1 000 cu ft) at standard pressure. Adjusting for actual pressure, the approximate moisture content is

Table 6.12 Moisture content of air for air temperatures from −80 to 40°C (ASHRAE, 1981, and Stover *et al.*, 1986)

Air Temperature		Moisture content, g/m³	Moisture content, lb/1 000 cu ft
°C	°F		
−80	−112	0.000 38	0.000 02
−75	−103	0.000 93	0.000 06
−70	−94	0.001 98	0.000 12
−65	−85	0.004 03	0.000 25
−60	−76	0.008 12	0.000 51
−55	−67	0.015 80	0.000 99
−50	−58	0.029 69	0.001 85
−45	−49	0.053 67	0.003 35
−40	−40	0.095 50	0.005 96
−35	−31	0.166 41	0.010 39
−30	−22	0.282 45	0.017 64
−25	−13	0.470 32	0.029 37
−20	−4	0.811 25	0.050 67
−15	5	1.229 08	0.076 76
−10	14	1.935 85	0.120 90
−5	23	2.994 24	0.187 00
0	32	4.565 02	0.285 10
5	41	6.532 90	0.408 00
10	50	9.227 72	0.576 30
15	59	12.881 65	0.804 50
20	68	17.773 32	1.110 00
25	77	24.291 81	1.517 10
30	86	32.907 86	2.055 20
35	95	44.260 37	2.764 20
40	104	59.178 75	3.695 90

$$17.8 \ g/m^3 \times 101.3 \ kPa/182 \ kPa$$

$$= 9.9 \ g/m^3 \quad (0.617 \ lb/1 \ 000 \ cu \ ft) \tag{6.9}$$

This represents the maximum amount of moisture that the air can hold at the given conditions of 20°C (68°F) and 182 kPa (26.43 psia).

Moisture loading to the desiccant dryer is, therefore,

$$9.9 \ g/m^3 \times 25.5 \ m^3/min = 252.5 \ g/min, \ or \ 15.1 \ kg/h$$

Next, the amount of desiccant required to dry the gas must be calculated:

$$Desiccant/(towers \ in \ operation) = 17.6 \ kg/kg \times 15.1 \ kg/h \times 16 \ h$$

$$= 4 \ 252 \ kg/towers \ that \ are \ drying \ (9 \ 567 \ lb/towers) \tag{6.10}$$

If three dryers are chosen, each with two towers, then the amount of desiccant required per tower is

$$Desiccant/tower = 4 \ 252 \ kg/3 \ dryers = 1 \ 417 \ kg/tower \ (3 \ 118 \ lb/tower)$$

The amount of desiccant for each dryer is twice this because each dryer has two towers, or 2 834 kg (6 234 lb). Each dryer will operate with only one tower drying at a time.

Similar calculations could also be performed when a refrigerant dryer rather than an aftercooler precedes the desiccant dryer.

This example illustrates some of the calculations required when designing an ozone disinfection system. Piping, offgas destruction, cooling, instrumentation, and process control are some of the other features that will need to be designed.

Current design procedures, as illustrated in this example, are based on survival data of coliform bacteria. It has been shown that a different approach may be required when viruses and protozoa are involved. If viruses and protozoa are the target organisms, then the advice of an expert in these areas is recommended.

OPERATION, MAINTENANCE, AND SAFETY

OPERATION. General. Operation is particularly important with respect to ozonation systems. Ozone generation is electrical power intensive; therefore, optimization of the system can significantly reduce operating costs (Rakness *et al.,* 1994). Constant attention must be given to the ozonation system to ensure that every watt of power applied to the system results in the maximum amount of ozone being transferred to the liquid under treatment. The ozonation system includes the following subunits: gas preparation, electrical power transformation, ozone generation, ozone diffusion, liquid- and gas-phase monitoring, and offgas destruction. The operator's contribution must include regular monitoring of the system to ensure that all subunits are fully operational. For example, the appropriate subunits should be receiving the necessary flow of coolant. The operator should continue to evaluate the process to determine whether the original system or initial mode of operation can be improved, for example, whether additions or modifications to the gas-preparation system would be required to ensure delivery of drier or cleaner gas to the generator. Modification to the mode of operation would include changing the point of application of the process gas.

Under the conventional construction bid format, it is difficult to ensure that the original design will be optimized for the system selected by the contractor. Optimization of the ozonation system can commence only after the equipment has been selected and the system placed in operation. Few wastewater treatment processes are as site-specific as ozonation.

Key items in operating an ozonation facility include the following:

- Delivery of a clean feed gas with a dew point below –60°C to the ozone generator;
- Maintenance of the required flow of generator coolant (air, water, or other liquid) medium;
- Operation of the ozone generator within its design parameters;
- Intimate contact of the required amount of process gas with the liquid under treatment; and
- Maintenance of ambient levels of ozone below the limits of applicable safety regulations.

Gas Preparation. A formalized air-preparation system-monitoring program, on a maximum 2-hour interval, must be established. Measurements to be obtained, where appropriate to the installed gas-preparation system, are as follows:

- Temperature and pressure profile of the system from gas inlet to discharge to the ozone generator;
- Temperature of coolant to and from devices such as aftercoolers and refrigerant dryers;
- Noise, vibration, and temperature of the pressurization unit along with other feasible physical measurements requiring more sophisticated monitoring devices;
- Noise, vibration, and temperature of the refrigerant drier; and
- Clogging or degeneration of the gas filter media.

The most critical parameter for measurement in the gas-preparation system is the point immediately before the ozone generator. Because a high dew point will result in lower ozone production, the operator must strive to operate the gas-preparation system at its maximum efficiency. The dew point must be regularly monitored in terms of both absolute number and trend to determine that the highest dew point does not exceed approximately –60°C (–76°F).

Electrical Power Supply. Operation of the electrical power supply typically suffers from a lack of monitoring equipment. Noise and temperature may be the only operational insights possible for the electrical power supply subunit.

Ozone Generator. Operation of the ozone generator involves controlling the applied voltage frequency to control the ozone thereby generated. The gas quantity and ozone concentration of the gas being discharged from the generator should be carefully monitored and recorded to determine the actual mass of ozone being generated.

The cooling media (water, air, or other material) must be carefully monitored for the ozone generator. Failure to maintain the required temperature levels in the ozone generator units will reduce production and may result in damage to the units.

Ozone Contactor. The ozone contactor and offgas-destruction system require monitoring to ensure efficiency of the contactor and personnel safety. Prolonged operational experience may further optimize the system. For example, in a multistage system contactor, the initial mode of operation may subsequently be modified to require a lower ozone dose for the same degree of disinfection by changing the amount of ozone injected into each stage. It may be most efficient to inject all of the ozone in the first stage, inject an equal amount throughout all of the stages, or use some other combination.

Offgas-Destruction System. The offgas-destruction system must be checked continually to maintain optimum conditions so that the levels of ozone gas discharged from the system are not exceeded.

Additional Considerations. The operator is charged with achieving the required level of disinfection for a given system. In its current state of evolution, wastewater ozonation requires that operators have a high level of insight into the problems of the specific systems for which they are responsible. Therefore, the operator must be alert to factors that may be affecting the ability of the facility to achieve the required levels of disinfection. These may include one or more of the factors previously discussed in this chapter. The operator must develop operational data that can justify physical modifications to the facility if disinfection results are poor. Data such as ozone use, ozone production, dew point, and other factors are required to justify further capital investments in the facility.

The operator must be alert for possible deterioration of system components. Ozone is a powerful oxidant and will vigorously attack materials not resistant to it. Despite the best efforts of all concerned with the design and construction of the ozonation system, faulty materials may be incorporated into the work with resulting hazards to operation and maintenance personnel. These material problems should be brought to the attention of maintenance personnel either for repair or for replacement with material that is more ozone resistant.

Periodic testing of the work spaces, as well as the overall WWTP area, should be carried out to ensure that the ambient levels of ozone concentration do not violate applicable health regulations.

MAINTENANCE. General. Regular maintenance by trained personnel with a variety of skills is essential for successful operation of the ozonation system. The various system components include the following:

- Gas preparation,
- Electric power supply,
- Ozone generator,
- Ozone contacting, and
- Associated equipment.

Gas Preparation. The gas-preparation system generally consists of process equipment that is familiar to maintenance, including

- Gas filters;
- Gas pressurization units—fan, blowers, and compressors;
- Aftercooler;
- Refrigerant drier;
- Desiccant dryers—pressure swing or thermally regenerated; and
- Oil coalescers.

Replacement cycles for the gas filter media are site-specific, depending on raw gas purity and the gas flow, but inspection intervals for the filters should not exceed 3 months. Although a generous supply of media should be provided for in the initial budget, operating experience will ultimately determine the site-specific media replacement schedule.

Gas-pressurization equipment is common to WWTPs and would be well within the normal experience of plant maintenance personnel. The units should be checked regularly to minimize the level of contamination of the feed gas by the pressurization device.

Shell and tube aftercoolers used to cool water require annual inspection for scaling or metal deterioration resulting from the passage of the cooling water through the heat exchanger. Cleaning or replacement would be based on visual inspection and other evaluation procedures.

Refrigerant dryers require personnel skilled in the operation and maintenance of air-conditioning or refrigeration equipment. Inspections at maximum 3-month intervals should be carried out by WWTP or service contractor personnel with the required skills and equipment. Minimum checks would include compressor belt tension and refrigerant pressure measurements.

Desiccant dryers warrant special attention by maintenance personnel, although specialized skills are not required. Special attention should be paid to the switch valve that directs gas to one or the other of the desiccant towers. Another key item is the device that controls the cycling periods of the drier. Weekly inspections of the unit should be performed to ensure proper operation of the unit. Maintenance is important to keep the unit in operation and prevent damage to the unit by fires, which can occur if a tower should continue for a prolonged period in the regeneration mode. Annual maintenance should include disassembly of the unit and inspection of the desiccant. The media should last approximately 10 years but may have to be replaced within a shorter period if the drier has been overloaded or poorly maintained.

Electrical Power Supply. The electrical power supply subunit should be carefully inspected and tested annually by specialized personnel. The efficiency of the various components of the subunits, control circuits, transformers, and associated electrical connections should be checked and tested to maintain the maximum efficiency of these systems. It may be necessary to obtain the services of specialized firms for this type of inspection and maintenance work.

Ozone Generator. The ozone generator is an unfamiliar unit to maintenance personnel and may justify, at least initially, entry into a maintenance contract with the generator supplier for emergency assistance and annual service. Specific personnel should be assigned to generator maintenance to allow them to gain sufficient familiarity and skills needed to perform tasks such as fuse replacement and dielectric tube cleaning, which will result in lower energy costs for a given ozone output. Because dielectric cleaning necessitates manpower and inevitable breakage and/or damage to the dielectrics during the cleaning process, the interval of cleaning should be optimized.

Ozone Contactor. The ozone contactor must be inspected annually because this is normally the only direct evaluation of the subunit. The chambers of the contactors should be well ventilated before entry by WWTP personnel.

Inspection of the contactor should address the structural as well as the functional integrity of the components of the contactor. Piping, valves, fittings, supports, brackets, and hatchways should be checked for deterioration that results from exposure to a highly oxidizing environment. Deteriorated material should be replaced. Proper records, including photographs, of the conditions observed during the annual inspection should be maintained.

The ozone-transfer efficiency of the gas diffusers in the contactor is a key factor in the operational cost of the facility. Fine-bubble diffusers are often used to achieve high ozone-transfer-system efficiencies, thereby decreasing the amount of ozone required and the associated power cost. The diffusion pattern of the ozone diffusers should be observed with the chamber partially filled with liquid. Treated, ozone-free feed gas should be used to test the diffuser system for any plugging or failure. Diffuser problems identified in the inspection should be corrected before placing the contactor back in operation.

The ozone-destruction unit, including enclosure, catalyst, and heating elements, should be completely inspected. The cost of replacing the catalyst is a considerable expenditure that can be deferred by good maintenance. Operational modes may be modified, based on observed deterioration, to prolong the catalyst life. Based on observed operational problems, several wastewater ozonation facilities have had to physically modify the ozone-destruction unit.

Associated Equipment. The instrumentation and controls of the ozonation system must be maintained regularly to optimize system operation. Flow meters, temperature and pressure sensors, and ozone analyzers must be kept in good condition so that the operator can measure the various parameters required for efficient operation.

SAFETY. Operation and maintenance personnel must be familiar with the hazards associated with ozone gas generation, conveyance, and destruction. Additional safety considerations that are involved if high-purity oxygen or recycled high-purity oxygen is used as the feed gas are not addressed in this manual.

The American Conference of Governmental Industrial Hygienists (1981) has recommended a threshold limit value for ozone of 0.2 mg/m^3 (0.1 ppm by volume at 25°C and 100 kPa pressure [77°F and 760 mm Hg]). ASTM Standard E591-77 has established the following standards for occupational exposure (ASTM, 1990):

- Eight-hour work day—0.2 mg/m^3 (0.1 ppm) time-weighted average, as measured by a minimum sampling time of 10 minutes; and
- Maximum concentration—0.6 mg/m^3 (0.3 ppm), as measured by a maximum sampling time of 10 minutes.

These standards are in a state of evolution and may change as further operational experience becomes available.

In addition to continued monitoring for ozone in the workspace, ASTM Standard E591-77 includes the following medical standards for personnel involved in ozonation facility operation:

- Comprehensive preplacement medical and occupational history,
- Preplacement physical examination with special emphasis on pulmonary condition and problems, and
- Periodic follow-up medical examinations.

Personnel exposed to ozone for prolonged periods of time may become accustomed to it and be unable to detect hazardous ozone levels. Therefore, properly calibrated ozone-monitoring equipment should be maintained in the work areas. These units should be set to sound an alarm at 0.2 mg/m^3 (0.1 ppm). They should shut down the ozone generators and start emergency ventilation systems at 0.6-mg/m^3 (0.3-ppm) ozone levels. Work areas should be ventilated for a minimum period of 10 minutes after a 0.6-mg/m^3 system alarm. The monitoring equipment should be tested weekly and recalibrated when necessary.

Ozone gas has been known to collect and concentrate in the generation equipment gas-distribution system for prolonged periods of time (LePage, 1979). Therefore, ozone generators and associated piping systems should be purged before opening the units or breaking the piping system. A minimum of two self-contained breathing apparatus should be maintained in the immediate vicinity of the ozonation system.

Entry into the ozone contactor should be carefully planned and carried out for maximum personnel safety. Personnel should be particularly aware of possible oxygen deficiencies at the bottom of the enclosed tanks even after efforts to purge the system. Steps to be followed include the following:

- Halt ozone flow to the contactor train. Valve off or, preferably, blank off the gas pipe to the contactor if possible.
- Purge the contactor with oxygen or prepared air.
- Valve off or, preferably, blank off all possible cross-connections with other ozonation trains.
- Open all access hatches and begin ventilation as dewatering of the contactor is carried out.
- Dewater the contactor.
- Continue the ventilation procedure.
- Have the contactor interior tested for ozone, oxygen, and gaseous contaminants such as hydrogen sulfide by personnel wearing protective clothing and self-contained breathing apparatus.
- Once working conditions have been determined to be safe, allow personnel equipped with a safety harness and rope to enter the contactors, using the buddy system. This entails each worker being tended by a second person on the deck of the contactor. Self-contained breathing apparatus are to be maintained immediately adjacent to the work area.

REFERENCES

American Conference of Governmental Industrial Hygienists (1981) *Documentation of the Threshold Limit Values.* 4th Ed., Cincinnati, Ohio.

American Public Health Association (1995) *Standard Methods for the Examination of Water and Wastewater.* 19th Ed., Washington, D.C.

American Society of Heating, Refrigerating, and Air Conditioning Engineers (1981) *Handbook: Fundamentals.* Atlanta, Ga.

Arthur, J.W., *et al.* (1975) *Comparative Toxicity of Sewage Effluent Disinfection to Freshwater Aquatic Life.* EPA-600/3-75-012, Ecol. Res. Ser., U.S. EPA, Cincinnati, Ohio.

ASTM (1990) *Water and Environmental Technology.* Philadelphia, Pa.

Bader, H., and Hoigné, J. (1981) Determination of Ozone in Water by the Indigo Method. *Water Res.* (G.B.), **15**, 449.

Bader, H., and Hoigné, J. (1982) Determination of Ozone in Water by the Indigo Method: A Submitted Standard Method. *Ozone Sci. Eng.*, **4**, 169.

Bader, H., *et al.* (1988) Photometric Method for the Determination of Low Concentrations of Hydrogen Peroxide by the Peroxidase Catalyzed Oxidation of N,N-Diethyl-*p*-Phenylenediamine (DPD). *Water Res.* (G.B.), **22**, 1109.

Bancroft, K., *et al.* (1984) Ozonation and Oxidation Competition Values. *Water Res.* (G.B.), **18**, 473.

Bell, J.B., and Smith, D.W. (1982) Wastewater Disinfection: Evaluation of Significant Factors. *Environ. Technol. Lett.*, **3**, 319.

Birdsall, C.M., *et al.* (1952) Iodometric Determination of Ozone. *Anal. Chem.*, **24**, 662.

Bollyky, L.J. (1979) Ozone: Safety and Health Considerations. *Proc. Seminar Des. Oper. Drinking Water Facilities Using Ozone or Chlorine Dioxide.* New Engl. Water Works Assoc., Dedham, Mass.

Bollyky, L.J., and Siegal, B. (1977) Disinfection of Secondary Effluent by Ozone. In *Forum on Ozone Disinfection.* E.G. Fochtman *et al.* (Eds.), Int. Ozone Inst., Cleveland, Ohio, 216.

Caverson, D.L., *et al.* (1986) Ozone Disinfection of Primary Effluent Using a Stirred Tank Reactor. *Can. J. Civ. Eng.*, **13**, 510.

Chang, S.D., and Singer, P.C. (1991) The Impact of Ozonation on Particle Stability and the Removal of TOC and THM Precursors. *J. Am. Water Works Assoc.*, **83**, 3, 71.

Chelkowska, K., *et al.* (1992) Numerical Simulations of Aqueous Ozone Decomposition: Comparison of Mechanisms. *Ozone Sci. Eng.*, **14**, 33.

Chheda, P., *et al.* (1992) Impact of Ozone on Stability of Montmorillonite Suspensions. *J. Coll. Inter. Sci.*, **153**, 226.

Chick, H. (1908) An Investigation of the Laws of Disinfection. *J. Hyg.* (G.B.), **8**, 92.

Chrostowski, P.C., *et al.* (1982) Laboratory Testing of Ozonation Systems Prior to Pilot-Plant Operations. *J. Am. Water Works Assoc.*, **74**, 38.

Coin, L., *et al.* (1964) Inactivation par l'Ozone du Virus de la Poliomyélite Présent dans les Eaux. *La Presse Médicale*, **72**, 2153.

Coin, L., *et al.* (1967) Inactivation par l'Ozone du Virus de la Poliomyélite Présent dans les Eaux. *La Presse Médicale*, **75**, 1883.

Corless, C., *et al.* (1989) Aqueous Ozonation of Quaternary Ammonium Surfactant. *Water Res.* (G.B.), **23**(11), 1367.

Dahi, E. (1976) Physicochemical Aspects of Disinfection of Water by Means of Ultrasound and Ozone. *Water Res.* (G.B.), **10**, 677.

Damez, F. (1982) Materials Resistant to Corrosion and Degradation in Contact With Ozone. In *Ozonization Manual for Water and Wastewater Treatment*. W.J. Masschelein (Ed.), John Wiley and Sons, Inc., New York, N.Y.

Danckwerts, P. (1970) *Gas Liquid Reactions.* McGraw-Hill, Inc., New York, N.Y.

Dawson, M., *et al.* (1974) Inactivation of Paralytic Shellfish Poison by Ozone Treatment. In *Fourth Food-Drugs from the Sea Meeting.* Mayaguez, Puerto Rico.

Diaper, E.W.J. (1972) Practical Aspects of Water and Waste Water Treatment by Ozone. In *Ozone in Water and Wastewater Treatment.* F.L. Evans (Ed.), Ann Arbor Science, Ann Arbor, Mich., 145.

Doré, M., *et al.* (1987) The Role of Alkalinity in the Effectiveness of Processes of Oxidation by Ozone. In *The Role of Ozone in Water and Wastewater Treatment, Proceedings of the 2nd International Conference.* D.W. Smith and G.R. Finch (Eds.), TekTran International, Kitchener, Ont., Can., 1.

Duguet, J.P., *et al.* (1989) Efficacy of the Combined Use of Ozone/UV or Ozone/Hydrogen Peroxide for Water Disinfection. *Proc. IOA Symp. Wasser Berlin '89, April 10–16.* Int. Ozone Assoc., Zürich, Switz., V5.1.

Edwards, M., and Benjamin, M.M. (1991) A Mechanistic Study of Ozone-Induced Particle Destabilization. *J. Am. Water Works Assoc.*, **83**, 6, 96.

Evison, L.M. (1978) Inactivation of Enteroviruses and Coliphages with Ozone in Water and Waste Waters. *Prog. Water Technol.*, **10**, 365.

Fair, G.M., *et al.* (1947) The Dynamics of Water Chlorination. *J. New Engl. Water Works Assoc.*, **61**, 285.

Fair, G.M., *et al.* (1948) The Behavior of Chlorine as a Water Disinfectant. *J. Am. Water Works Assoc.*, **40**, 1051.

Farooq, S. (1976) Criteria of Design of Ozone Disinfection Plants. In *Forum on Ozone Disinfection.* E.G. Fochtman *et al.* (Eds.), Int. Ozone Inst., Chicago, Ill., 394.

Feachem, R.G., *et al.* (1983) *Sanitation and Disease: Health Aspects of Excreta and Wastewater Management.* John Wiley & Sons, Inc., New York, N.Y.

Finch, G.R., and Fairbairn, N. (1991) Comparative Inactivation of Poliovirus Type 3 and MS2 Coliphage in Demand-Free Buffer Using Ozone. *Appl. Environ. Microbiol.*, **57**, 3121.

Finch, G.R., and Smith, D.W. (1989) Ozone Dose-Response of *Escherichia coli* in Activated Sludge Effluent. *Water Res.* (G.B.), **23**, 1017.

Finch, G.R., and Smith, D.W. (1990) Evaluation of Empirical Process Design Relationships for Ozone Disinfection of Water and Wastewater. *Ozone Sci. Eng.*, **12**, 157.

Finch, G.R., and Smith, D.W. (1991) Pilot-Scale Evaluation of the Effects of Mixing on Ozone Disinfection of *Escherichia coli* in a Semi-Batch Stirred Tank Reactor. *Ozone Sci. Eng.*, **13**, 593.

Finch, G.R., *et al.* (1988) Dose-Response of *Escherichia coli* in Ozone Demand-Free Phosphate Buffer. *Water Res.* (G.B.), **22**, 1563.

Finch, G.R., *et al.* (1992) Inactivation of *Escherichia coli* Using Ozone and Ozone-Hydrogen Peroxide. *Environ. Technol.* (G.B.), **13**, 571.

Finch, G.R., *et al.* (1993a) Comparison of *Giardia lamblia* and *Giardia muris* Cyst Inactivation by Ozone. *Appl. Environ. Microbiol.*, **59**, 3674.

Finch, G.R., *et al.* (1993b) Ozone Inactivation of *Cryptosporidium parvum* in Demand-Free Phosphate Buffer Determined by In Vitro Excystation and Animal Infectivity. *Appl. Environ. Microbiol.*, **59**, 4203.

Finch, G.R., *et al.* (1994) *Ozone Disinfection of* Giardia *and* Cryptosporidium. Am. Water Works Assoc. Res. Found.; Am. Water Works Assoc., Denver, Colo.

Ghan, H.B., *et al.* (1977) The Significance of Water Quality on Wastewater Disinfection with Ozone. In *Forum on Ozone Disinfection*. E.G. Fochtman *et al.* (Eds.), Int. Ozone Inst., Cleveland, Ohio, 46.

Given, P.W., and Smith, D.W. (1979) Disinfection of Dilute, Low Temperature Wastewater Using Ozone. *Ozone Sci. Eng.*, **1**, 91.

Glaze, W.H. (1986) Reaction Products of Ozone: A Review. *Environ. Health Perspective*, **69**, 151.

Gordon, G., *et al.* (1992) *Disinfectant Residual Measurement Methods*. 2nd Ed., Am. Water Works Assoc., Denver, Colo.

Graham, N.J.D., *et al.* (1992) Alternative Disinfection Regimes for Trihalomethane Control—Significance of Pre-Disinfectant Dose. *Environ. Technol.* (G.B.), **13**, 461.

Grasso, D., *et al.* (1990) Ozone Mass Transfer in a Gas-Sparged Turbine Reactor. *Res. J. Water Pollut. Control Fed.*, **62**, 246.

Grunwell, J., *et al.* (1986) A Detailed Comparison of Analytical Methods for Residual Ozone Measurement. In *Analytical Aspects of Ozone Treatment of Water and Wastewater*. R.G. Rice *et al.* (Eds.), Lewis Publishers, Inc., Chelsea, Mich., 91.

Guittonneau, S., *et al.* (1992) Characterization of Natural Water for Potential to Oxidize Organic Pollutants with Ozone. *Ozone Sci. Eng.*, **14**, 185.

Haas, C.N., and Joffe, J. (1994) Disinfection Under Dynamic Conditions—Modification of Hom Model for Decay. *Environ. Sci. Technol.*, **28,** 1367.

Haas, C.N., and Karra, S.B. (1984) Kinetics of Microbial Inactivation by Chlorine. I. Review of Results in Demand-Free Systems. *Water Res. (G.B.)*, **18,** 1443.

Haas, C.N., *et al.* (1993) *Experimental Methodologies for the Determination of Disinfection Effectiveness.* Am. Water Works Assoc. Res. Found.; Am. Water Works Assoc., Denver, Colo.

Haas, C.N., *et al.* (1994) *Development and Validation of Rational Design Methods of Disinfection.* Am. Water Works Assoc.; Am. Water Works Assoc. Res. Found., Denver, Colo.

Hamelin, C., and Chung, Y.S. (1974) Optimal Conditions for Mutagenesis by Ozone in *Escherichia coli* K12. *Mutat. Res.*, **24,** 271.

Hamelin, C., *et al.* (1977) Ozone-Induced DNA Degradation in Different DNA Polymerase I Mutants of *Escherichia coli* K12. *Biochem. Biophys. Res. Commun.*, **77,** 220.

Hamon, J.L. (1982) Determination of Residual Ozone in Water. In *Ozonization Manual for Water and Wastewater Treatment.* W.J. Masschelein (Ed.), John Wiley & Sons, Inc., New York, N.Y., 162.

Hann, V.A., and Manley, T.C. (1952) Ozone. In *Encyclopedia of Chemical Technology.* Vol. 9, John Wiley & Sons, Inc., New York, N.Y., 735.

Helmer, R.D. (1992) MS2 Coliphage and HPC Bacteria as Indicators of Ozone Disinfection Performance. M.S. thesis, Univ. of Alberta, Can.

Hoff, J.C. (1986) *Inactivation of Microbial Agents by Chemical Disinfectants.* EPA-600/2-86-067, U.S. EPA, Water Eng. Res. Lab., Cincinnati, Ohio.

Hoff, J.C. (1987) Strengths and Weaknesses of Using CT Values to Evaluate Disinfection Practice. In *AWWA Seminar on Assurance of Adequate Disinfection, or CT or Not CT.* Am. Water Works Assoc., Denver, Colo., 49.

Hoigné, J. (1982) Mechanisms, Rates and Selectivities of Oxidations of Organic Compounds Initiated by Ozonation of Water. In *Handbook of Ozone Technology and Applications.* Vol. 1, R.G. Rice and A. Netzer (Eds.), Ann Arbor Science, Ann Arbor, Mich., 341.

Hoigné, J. (1994) Characterization of Water Quality Criteria for Ozonation Processes. Part I: Minimal Set of Analytical Data. *Ozone Sci. Eng.*, **16,** 113.

Hoigné, J., and Bader, H. (1975) Identification and Kinetic Properties of the Oxidizing Decomposition Products of Ozone in Water and Its Impact on Water Purification. *2nd Int. Symp. Ozone Technol., Montreal, Can.*, Ozone Press Int., Jamesville, N.Y., 271.

Hoigné, J., and Bader, H. (1978) Ozonation of Water: Kinetics of Oxidation of Ammonia by Ozone and Hydroxyl Radicals. *Environ. Sci. Technol.*, **12,** 79.

Hoigné, J., and Bader, H. (1983a) Rate Constants of Reactions of Ozone with Organic and Inorganic Compounds in Water. I. Non-Dissociating Organic Compounds. *Water Res.* (G.B.), **17**, 173.

Hoigné, J., and Bader, H. (1983b) Rate Constants of Reactions of Ozone with Organic and Inorganic Compounds in Water. II. Dissociating Organic Compounds. *Water Res.* (G.B.), **17**, 185.

Hoigné, J., and Bader, H. (1994) Characterization of Water Quality Criteria for Ozonation Processes. Part II: Lifetime of Added Ozone. *Ozone Sci. Eng.*, **16**, 121.

Hoigné, J., *et al.* (1985) Rate Constants of Reactions of Ozone with Organic and Inorganic Compounds in Water. III. Inorganic Compounds and Radicals. *Water Res.* (G.B.), **19**, 993.

Holluta, J., and Unger, U. (1954) The Destruction of *Escherichia coli* by Chlorine Dioxide and Ozone. *Vom Wasser* (Ger.), **21**, 129.

Huck, P.M., *et al.* (1992) Biodegradation of Aquatic Organic Matter with Reference to Drinking Water Treatment. *Sci. Total. Environ.*, **118**, 531.

International Ozone Association (1987a) Calibration Method for Residual Ozone by the Oxidation of Nitrite. Rep. 003/87, Stand. Committee—Eur., Lille, Fr.

International Ozone Association (1987b) Colorimetric Procedure for the Determination of Residual Ozone in Water (ACVK—Method). Rep. 005/87, Stand. Committee—Eur., Lille, Fr.

International Ozone Association (1987c) Colorimetric Procedure for the Determination of Residual Ozone in Water (Indigo Trisulphonate—Method). Rep. 004/87, Stand. Committee—Eur., Lille, Fr.

International Ozone Association (1987d) Colorimetric Procedure for the Determination of Traces of Ozone in Water (Indigo Trisulphonate—Method). Rep. 006/87, Stand. Committee—Eur., Lille, Fr.

International Ozone Association (1987e) Electrochemical Method for Continuous Measurement of Residual Ozone in Water. Rep. 007/87, Stand. Committee—Eur., Lille, Fr.

International Ozone Association (1987f) Ozone Concentration Measurement in a Process Gas by U.V. Absorption. Rep. 002/87, Stand. Committee—Eur., Lille, Fr.

International Ozone Association (1987g) Photometric Measurement of Low Ozone Concentrations in the Gas Phase. Rep. 008/87, Stand. Committee—Eur., Lille, Fr.

Ishizaki, K., *et al.* (1987) Effect of Ozone on Plasmid DNA of *Escherichia coli In Situ. Water Res.* (G.B.), **21**(7), 823.

Jakubowski, W. (1990) The Control of *Giardia* in Water Supplies. In *Giardiasis*. Vol. 3, E.A. Meyer (Ed.), Elsevier, Amsterdam, Neth., 335.

James M. Montgomery Consulting Engineers (1989) *Design Report for Oxidation Demonstration Project.* Metro. Water Dist. South. Calif., Pasadena, Calif.

Jarroll, E.L., Jr. (1988) Effect of Disinfectants on *Giardia* Cysts. *Crit. Rev. Environ. Control*, **18**, 1.

Jekel, M.R. (1994) Flocculation Effects of Ozone. *Ozone Sci. Eng.*, **16**, 55.

Katzenelson, E., *et al.* (1974) Inactivation Kinetics of Viruses and Bacteria by Use of Ozone. *J. Am. Water Works Assoc.*, **66**, 725.

Kinman, R.N. (1975) Water and Wastewater Disinfection with Ozone: A Critical Review. *Crit. Rev. Environ. Control*, **5**, 141.

Korich, D.G., *et al.* (1990) Effects of Ozone, Chlorine Dioxide, Chlorine, and Monochloramine on *Cryptosporidium parvum* Oocyst Viability. *Appl. Environ. Microbiol.*, **56**, 1423.

Kott, Y., *et al.* (1978) Coliphage Survival as Viral Indicators in Various Wastewater Quality Effluents. *Prog. Water Technol.*, **10**, 337.

Krasner, S.W., *et al.* (1993) Formation and Control of Bromate During Ozonation of Waters Containing Bromide. *J. Am. Water Works Assoc.*, **85**, 73.

Labatiuk, C.W., *et al.* (1992) Factors Influencing the Infectivity of *Giardia muris* Cysts Following Ozone Inactivation in Laboratory and Natural Waters. *Water Res.* (G.B.), **26**, 733.

Labatiuk, C.W., *et al.* (1994) Inactivation of *Giardia muris* Using Ozone and Ozone-Hydrogen Peroxide. *Ozone Sci. Eng.*, **16**, 67.

Langlais, B., *et al.* (1989) Improvement of a Biological Treatment by Prior Ozonation. *Ozone Sci. Eng.*, **11**, 155.

Langlais, B., *et al.* (1990) The C.T. Value Concept for Evaluation of Disinfection Process Efficiency; Particular Case of Ozonation for Inactivation of Some Protozoan: Free Living Amoeba and *Cryptosporidium*. In *New Developments: Ozone in Water and Wastewater Treatment. Proc. Int. Ozone Assoc. Spring Conf.,* Shreveport, La.

Langlais, B., *et al.* (Eds.) (1991) *Ozone in Water Treatment: Application and Engineering.* Lewis Publishers, Inc., Chelsea, Mich.

LePage, W.L. (1979) A Plant Operator's View of Ozonation. *Proc. Seminar Des. Oper. Drinking Water Facilities Using Ozone or Chlorine Dioxide.* New Engl. Wastewater Assoc. (Ed.), Dedham, Mass.

Lev, O., and Regli, S. (1992a) Evaluation of Ozone Disinfection Systems: Characteristic Concentration C. *J. Environ. Eng.*, **118**, 477.

Lev, O., and Regli, S. (1992b) Evaluation of Ozone Disinfection Systems: Characteristic Time T. *J. Environ. Eng.*, **118**, 268.

Lorenzo-Lorenzo, M.J., *et al.* (1993) Effect of Ultraviolet Disinfection of Drinking Water on the Viability of *Cryptosporidium parvum* Oocysts. *J. Parasitol.*, **79**, 67.

Maclean, S., *et al.* (1975) Effects of Ozone-Treated Seawater on the Spawned, Fertilized, Meiotic, and Cleaving Eggs of the Commercial American Oyster. *Mutat. Res.*, **21**, 283.

Maggiolo, A. (1978) Ozone's Radical and Ionic Mechanisms of Reaction with Organic Compounds in Water. In *Ozone/Chlorine Dioxide Oxidation*

Products of Organic Materials. R.G. Rice and J.A. Cotruvo (Eds.), Ozone Press Int., Cleveland, Ohio, 59.

Maier, D. (1984) Microflocculation by Ozone. In *Handbook of Ozone Technology and Applications*. Vol. 2, R.G. Rice and A. Netzer (Eds.), Butterworth Publishers, Stoneham, Mass., 123.

Majumdar, B., and Sproul, O.J. (1974) Technical and Economic Aspects of Water and Wastewater Ozonation: A Critical Review. *Water Res.* (G.B.), **8**, 253.

Malcolm Pirnie, Inc., and HDR Engineering, Inc. (1991) *Guidance Manual for Compliance with the Filtration and Disinfection Requirements for Public Water Systems Using Surface Water Sources*. Am. Water Works Assoc., Denver, Colo.

Mangum, D.C., and McIlhenny, W.F. (1975) Control of Marine Fouling in Intake Systems–A Comparison of Ozone and Chlorine. In *Aquatic Applications of Ozone*. W.J. Blogoslawski and R.G. Rice (Eds.), International Ozone Institute, Inc., Syracuse, N.Y., 138.

Mariñas, B.J., *et al.* (1991) Performance of Ozone Bubble-Diffuser Contactors: A Pilot-Scale Study. *Proc. 1991 Annu. Conf.* Am. Water Works Assoc., Philadelphia, Pa.

Martin, N., *et al.* (1992) Design and Efficiency of Ozone Contactors for Disinfection. *Ozone Sci. Eng.*, **14**, 391.

Metropolitan Water District of Southern California and James M. Montgomery Consulting Engineers (1991) *Pilot-Scale Evaluation of Ozone and Peroxone*. Am. Water Works Assoc. Res. Found., Denver, Colo.

Nebel, C. (1981) Ozone. In *Encyclopedia of Chemical Technology*. Vol. 16, John Wiley & Sons, Inc., New York, N.Y., 684.

Novak, F., *et al.* (1977) Municipal Disinfection with Ozone and Without Filtration. In *Forum on Ozone Disinfection*. E.G. Fochtman *et al.* (Eds.), Int. Ozone Inst., Cleveland, Ohio, 98.

Ohlrogge, J.B., and Kernan, T.P. (1983) Toxicity of Activated Oxygen: Lack of Dependence on Membrane Fatty Acid Composition. *Biochem. Biophys. Res. Commun.*, **113**, 301.

Pavoni, J.L., and Tittlebaum, M.E. (1972) Virus Inactivation in Secondary Wastewater Treatment Plant Effluent Using Ozone. In *Virus Survival in Water and Wastewater Systems*. J.F. Malina and B.P. Sagik (Eds.), Univ. of Texas, Austin, 180.

Peeters, J.E., *et al.* (1989) Effect of Disinfection of Drinking Water with Ozone or Chlorine Dioxide on Survival of *Cryptosporidium parvum* Oocysts. *Appl. Environ. Microbiol.*, **55**, 1519.

Pichet, P., and Hurtubise, C. (1975) Reactions of Ozone in Artificial Seawater. In *Second International Symposium on Ozone Technology*. Montreal, Que., Can.

Pryor, W.A., *et al.* (1983) Mechanisms for the Reaction of Ozone with Biological Molecules: The Source of the Toxic Effects of Ozone. In *Advances in Modern Environmental Toxicology.* M.G. Mustafa and M.A. Mehlman (Eds.), Ann Arbor Science, Ann Arbor, Mich., 7.

Rakness, K.L., *et al.* (1984) Design, Start-Up, and Operation of an Ozone Disinfection Unit. *J. Water Pollut. Control Fed.*, **56**, 1152.

Rakness, K.L., *et al.* (1988) Practical Design Model for Calculating Bubble Diffuser Contactor Ozone Transfer Efficiency. *Ozone Sci. Eng.*, **10**, 173.

Rakness, K.L., *et al.* (1992) Operating Strategy to Meet SWTR Disinfection Regulations at the Los-Angeles-Aqueduct-Filtration-Plant. *Ozone Sci. Eng.*, **14**, 439.

Rakness, K.L., *et al.* (1993) Wastewater Disinfection with Ozone Process Control and Operating Results. *Ozone Sci. Eng.*, **15**, 497.

Rakness, K.L., *et al.* (1994) Ozone Contactor Sizing Based on Demand, Decay and Equipment Criteria. In *IOA, Pan American Committee Meeting.* Int. Ozone Assoc., Richmond, Va.

Rice, E.W., and Hoff, J.C. (1981) Inactivation of *Giardia lamblia* Cysts by Ultraviolet İrradiation. *Appl. Environ. Microbiol.*, **42**, 546.

Robson, C.M. (1982) Design Engineering Aspects of Ozonation Systems. In *Handbook of Ozone Technology and Applications.* Vol. 1, R.G. Rice and A. Netzer (Eds.), Ann Arbor Science, Ann Arbor, Mich., 143.

Rosen, H. (1973) Use of Ozone and Oxygen in Advanced Wastewater Treatment. *J. Water Pollut. Control Fed.*, **45**, 2521.

Rosen, H.M. (1976) Ozone Wastewater Disinfection: State-of-the-Art. In *Forum on Ozone Disinfection.* E.G. Fochtman et al. (Ed.), Int. Ozone Inst., Chicago, Ill., 36.

Ross, W.R., *et al.* (1976) Studies on Disinfection and Chemical Oxidation with Ozone and Chlorine in Water Reclamation. *Proc. 2nd Int. Symp. Ozone Technol.,* Int. Ozone Assoc., Montreal, Can.

Roustan, M., *et al.* (1987) Mass Balance Analysis of Ozone in Conventional Bubble Contactors. *Ozone Sci. Eng.*, **9**, 289.

Safe Drinking Water Committee (1980) *Drinking Water and Health.* National Academy Press, Washington, D.C.

Scaccia, C., and Rosen, H.M. (1978) Ozone Contacting: What Is the Answer? *Ozonews*, **5**, 10(II), 1.

Schmidtke, N.W., and Smith, D.W. (Eds.) (1983) Introduction to Scale-up of Water and Wastewater Treatment Processes. In *Scale-up of Water and Wastewater Treatment Processes.* Butterworth Publishers, Boston, Mass., 11.

Scott, D.B.M., and Lesher, E.C. (1963) Effect of Ozone on Survival and Permeability of *Escherichia coli. J. Bacteriol.*, **85**, 567.

Sengupta, C., *et al.* (1975) Power Plant Cooling Water Treatment with Ozone. In *Aquatic Applications of Ozone.* W.J. Blogoslawski and R.G. Rice (Eds.), Int. Ozone Inst., Inc., Syracuse, N.Y., 120.

Shuval, H.I., and Gruener, N. (1973) Health Considerations in Renovating Wastewater for Domestic Use. *Environ. Sci. Technol.*, **7**, 600.

Singer, P.C. (1994) Control of Disinfection By-Products in Drinking Water. *J. Environ. Eng.*, **120**, 727.

Singer, P.C., and Zilli, W.B. (1975) Ozonation of Ammonia in Wastewater. *Water Res.* (G.B.), **9**, 127.

Somich, C.J., *et al.* (1990) On-Site Treatment of Pesticide Waste and Rinsate Using Ozone and Biologically Active Soil. *Environ. Sci. Technol.*, **24**, 745.

Sproul, O.J., *et al.* (1979) *Effect of Particulates on Ozone Disinfection of Bacteria and Viruses in Water.* EPA-600/2-79-089, U.S. EPA, Cincinnati, Ohio.

Staehelin, J., and Hoigné, J. (1981) Zur Chemischen Reaktionskinetik des Ozonzerfalls in Wasser. *Wasser '81* (Ger.), **2**, 623.

Staehelin, J., and Hoigné, J. (1985) Decomposition of Ozone in Water in the Presence of Organic Solutes Acting as Promoters and Inhibitors of Radical Chain Reactions. *Environ. Sci. Technol.*, **19**, 1206.

Stankovic, I. (1991) Optimization of Ozone Contactors in a Water Treatment Plant Using Mass-Transfer Correlations. In *Chemistry for the Protection of the Environment, Environmental Science Research.* L. Pawlowski *et al.* (Eds.), Plenum Publishing Co., New York, N.Y., 583.

Stanley, J.H., and Johnson, J.D. (1979) Amperometric Membrane Electrode for Measurement of Ozone in Water. *Anal. Chem.*, **51**, 2144.

Stokinger, H.E. (1959) *Factors Modifying Toxicity of Ozone.* Advances in Chemistry Series #21, Am. Chem. Soc., Washington, D.C.

Stover, E.L., and Jarnis, R.N. (1980) Engineering and Economic Aspects of Wastewater Disinfection with Ozone Under Stringent Bacteriological Standards. *Ozone Sci. Eng.*, **2**, 159.

Stover, E.L., and Jarnis, R.N. (1981) Obtaining High-Level Wastewater Disinfection with Ozone. *J. Water Pollut. Control Fed.*, **53**, 1637.

Stover, E.L., *et al.* (1986) *Municipal Wastewater Disinfection.* EPA-625/1-86-021, U.S. EPA, Cincinnati, Ohio.

Stover, E.L., *et al.* (1987) Full-scale Testing of Four Different Ozone Contactors. *2nd Int. Conf. Role Ozone Water Wastewater Treat.* D.W. Smith and G.R. Finch (Eds.), Edmonton, Alb., Can., 321.

Symons, J.M., *et al.* (1994) Precursor Control in Waters Containing Bromide. *J. Am. Water Works Assoc.*, **86**, 6, 48.

Thurberg, F. (1975) Inactivation of Red-Tide by Ozone Treatment. In *Aquatic Applications of Ozone.* W.J. Blogoslawski and R.G. Rice (Eds.), Int. Ozone Inst., Inc., Syracuse, N.Y., 50.

Tobin, R.J. (1987) Indicator Systems for Microbiological Quality and Safety of Water. *J. Environ. Pathol. Toxicol. Oncol.*, **7**, 5/6, 115.

Tomiyasu, H., *et al.* (1985) Kinetics and Mechanism of Ozone Decomposition in Basic Aqueous Solution. *Inorg. Chem.*, **24**, 2962.

Trussell, R.R. (1992) Oxidation By-Products Complicate Disinfectant Choices. *Waterworld News*, **8**, 14.

U.S. Environmental Protection Agency (1989) Drinking Water; National Primary Drinking Water Regulations; Filtration, Disinfection; Turbidity, *Giardia lamblia*, Viruses, *Legionella*, and Heterotrophic Bacteria; Final Rule. *Fed. Reg.*, **54**, 27486.

Venosa, A.D. (1972) Ozone as a Water and Wastewater Disinfectant: Literature Review. In *Ozone in Water and Wastewater Treatment*. F.L. Evans (Ed.), Ann Arbor Science, Ann Arbor, Mich., 83.

Venosa, A.D., *et al.* (1978) Comparative Efficiences of Ozone Utilization and Microorganism Reduction in Different Ozone Contactors. In *Progress in Wastewater Disinfection Technology. Proc. Natl. Symp.,* A.D. Venosa (Ed.), EPA-600/9-79-018, U.S. EPA, Munic. Environ. Res. Lab., Cincinnati, Ohio, 144.

Venosa, A.D., *et al.* (1980) Disinfection of Filtered and Unfiltered Secondary Effluent in Two Ozone Contactors. *Environ. Int.*, **4**, 299.

Venosa, A.D., *et al.* (1985) Reliable Ozone Disinfection Using Off-Gas Control. *J. Water Pollut. Control Fed.*, **57**, 929.

Wallis, P.M., *et al.* (1990) Inactivation of *Giardia* Cysts in a Pilot Plant Using Chlorine Dioxide and Ozone. *Proc. Water Qual. Technol. Conf.,* Am. Water Works Assoc., Philadelphia, Pa., 695.

Ward, R.W. (1977) Effects of Residual Ozone on Aquatic Organisms. In *Forum on Ozone Disinfection*. E.G. Fochtman *et al.* (Eds.), Int. Ozone Inst., Cleveland, Ohio, 260.

Ward, R.W., *et al.* (1976) *Disinfection Efficiency and Residual Toxicity of Several Wastewater Disinfectants.* EPA-600/2-76-156, U.S. EPA, Cincinnati, Ohio.

Watson, H.E. (1908) A Note on the Variation of the Rate of Disinfection with Change in the Concentration of the Disinfectant. *J. Hyg.* (G.B.), **8**, 536.

Weast, R.C. (Ed.) (1987) *CRC Handbook of Chemistry and Physics.* 67th Ed., CRC Press, Inc., Boca Raton, Fla.

Wickramanayake, G.B. (1984) Kinetics and Mechanism of Ozone Inactivation of Protozoan Cysts. Ph.D. thesis, Ohio State Univ., Athens.

Wickramanayake, G.B., and Sproul, O.J. (1988) Ozone Concentration and Temperature Effects on Disinfection Kinetics. *Ozone Sci. Eng.*, **10**, 123.

Wickramanayake, G.B., *et al.* (1984a) Inactivation of *Giardia lamblia* Cysts with Ozone. *Appl. Environ. Microbiol.*, **48**, 671.

Wickramanayake, G.B., *et al.* (1984b) Inactivation of *Naegleria* and *Giardia* Cysts in Water by Ozonation. *J. Water Pollut. Control Fed.*, **56**, 983.

Wickramanayake, G.B., *et al.* (1985) Effects of Ozone and Storage Temperature on *Giardia* Cysts. *J. Am. Water Works Assoc.*, **77**, 8, 74.

Wolfe, R.L., *et al.* (1989a) Disinfection of Model Indicator Organisms in a Drinking Water Pilot Plant by Using Peroxone. *Appl. Environ. Microbiol.*, **55**, 2230.

Wolfe, R.L., *et al.* (1989b) Inactivation of *Giardia muris* and Indicator Organisms Seeded in Surface Water Supplies by Peroxone and Ozone. *Environ. Sci. Technol.*, **23**, 744.

Zhou, H., and Smith, D.W. (1994) Kinetics of Ozone Disinfection in a Completely Mixed System. *J. Environ. Eng.*, **120**, 841.

Zhou, H., *et al.* (1994) Modeling of Dissolved Ozone Concentration Profiles in Bubble Columns. *J. Environ. Eng.*, **120**, 821.

Chapter 7
Ultraviolet Disinfection

227 Introduction
230 General Description of Ultraviolet Disinfection
233 Photoreactivation and Dark Repair
236 Ultraviolet Inactivation Kinetics
244 Other Kinetic Models
248 Effect of Intensity on Inactivation Behavior
249 Intensity
257 Ultraviolet Dose
259 Hydraulics
259   Longitudinal Dispersion
264   Head Loss
267 Factors Affecting Lamp Output
269 Mathematical Models
273 Fouling
275 General Considerations in Ultraviolet System Design

276 Design Wastewater Characteristics
279 Pilot Testing
280 System Sizing and Configuration Considerations
282 Retrofit Considerations
282 Current Ultraviolet Equipment
283 Low-Pressure Mercury Lamp Systems
284 Horizontal Ultraviolet Systems
285 Vertical Ultraviolet Systems
286 Medium-Pressure Mercury Lamp Systems
288 Low-Pressure, High-Intensity Systems
289 References

*I*NTRODUCTION

Chlorination has been the de facto choice for most wastewater disinfection operations since the early 1900s. Although chlorination is still used in the majority of disinfection applications, alternative processes are increasingly being selected. Ultraviolet (UV) irradiation has become the most common alternative to chlorination for wastewater disinfection in North America.

The emergence of UV irradiation as an important wastewater disinfection alternative may be attributed to the drawbacks of conventional chlorination, improvements in UV technology, and advances in our understanding of the UV process. The major problems associated with chlorination are effluent toxicity and safety. Free and combined chlorine elicit a toxic response in fish and daphnids at extremely low concentrations (U.S. EPA, 1986b). Residual chlorine may be effectively eliminated by dechlorination (as is required in most new discharge permits), but effluent toxicity will remain in some cases (Rein *et al.*, 1992). Factors that may contribute to toxicity following chlorination/dechlorination include chlorinated disinfection byproducts, active (+1 valent) chlorine not removed by dechlorination (Helz and Nweke, 1995), and unoxidized ammonia (dechlorination restores chloramines to ammoniacal nitrogen).

Chlorine is usually applied in a gaseous form. Although few accidents have occurred with gaseous chlorine (White, 1992), it does represent a potential hazard to human health and the environment. As a result, the Uniform Fire Code has been amended such that containment and scrubbing facilities are required for gaseous chlorine application (see references in WEF, 1993).

Dechlorination and containment facility requirements have increased the cost of chlorine-based disinfection. At the same time, the development and application of open-channel, modular systems have reduced the cost of UV disinfection. Consequently, the costs of the two processes are comparable for new facilities (Putnam *et al.*, 1993).

Probably in response to these developments, the frequency with which UV has been selected for disinfection has increased in recent years. Among U.S. wastewater treatment plants (WWTPs), only approximately 50 used UV disinfection in 1986; most of these facilities had relatively small flows ($Q < 1$ mgd). By 1990, more than 500 WWTPs had adopted UV disinfection, a significant fraction of them at large facilities ($Q > 10$ mgd). Today, more than 1 000 WWTPs in North America have chosen UV irradiation for wastewater disinfection.

The majority of UV disinfection systems today employ an open-channel, modular design. Two principal lamp geometries have been adopted: horizontal, uniform arrays with flow directed parallel to lamp axes and vertical, staggered arrays with flow directed perpendicular to lamp axes (see Figure 7.1). The horizontal lamp orientation has been adopted in the majority of applications.

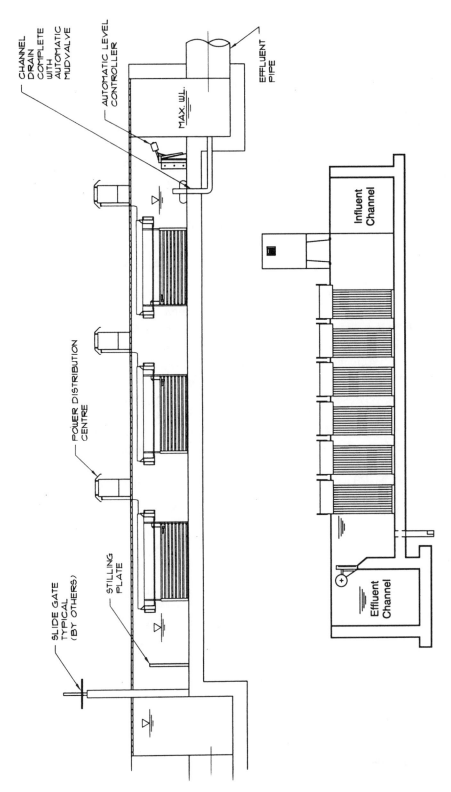

CHANNEL DRAIN COMPLETE WITH AUTOMATIC MUDVALVE

AUTOMATIC LEVEL CONTROLLER

MAX. WL.

EFFLUENT PIPE

POWER DISTRIBUTION CENTRE

SLIDE GATE TYPICAL (BY OTHERS)

STILLING PLATE

Influent Channel

Effluent Channel

Figure 7.1   Top, schematic illustration of open–channel ultraviolet disinfection system with horizontal lamp configuration; bottom, schematic illustration of open–channel ultraviolet disinfection system with vertical lamp configuration

GENERAL DESCRIPTION OF ULTRAVIOLET DISINFECTION

Ultraviolet irradiation is a physical disinfection process, and as such, it has several fundamental characteristics that distinguish it from chemical disinfection processes (such as chlorination). Ultraviolet irradiation achieves disinfection by inducing photobiochemical changes within microorganisms. At a minimum, two conditions must be met for a photochemical reaction to take place:

- Radiation of sufficient energy to alter chemical bonds must be available, and
- Such radiation must be absorbed by the target molecule (organism).

In the majority of UV disinfection applications, low-pressure mercury arc lamps have been chosen as the source of UV radiation. Approximately 85% of the output from these lamps is monochromatic at a wavelength (λ) of 253.7 nm (see Figure 7.2a). Several other lines are evident in the output spectrum from a low-pressure mercury arc lamp. A line that exists at 185 nm represents radiation with much higher energy than at 253.7 nm, but this line is unimportant in most applications because of absorbance by the quartz jackets surrounding the lamp and by aqueous constituents. Several small lines are evident in the visible range ($\lambda \geq 400$ nm). These radiation lines are ineffective with respect to disinfection, but are responsible for the pale blue color displayed by low-pressure lamps.

The energy associated with electromagnetic radiation may be calculated as

$$E_\lambda = \frac{hC}{\lambda} A \qquad (7.1)$$

Where

$\quad E_\lambda \quad = \quad$ radiant energy associated with given wavelength, kcal/einstein;

$\quad C \quad = \quad$ speed of electromagnetic radiation in a vacuum, 3.00×10^{17} nm/s;

$\quad h \quad = \quad$ Planck's constant = 1.583×10^{-37} kcal $\cdot$ s;

$\quad \lambda \quad = \quad$ wavelength of electromagnetic radiation, nm; and

$\quad A \quad = \quad$ Avogadro's number $\approx 6.023 \times 10^{23}$ photons/einstein.

(Note: In a photochemical reaction, one einstein represents one "mole" [Avogadro's number] of photons. Photochemical reactions almost always proceed via interactions between single photons and single molecules. Therefore,

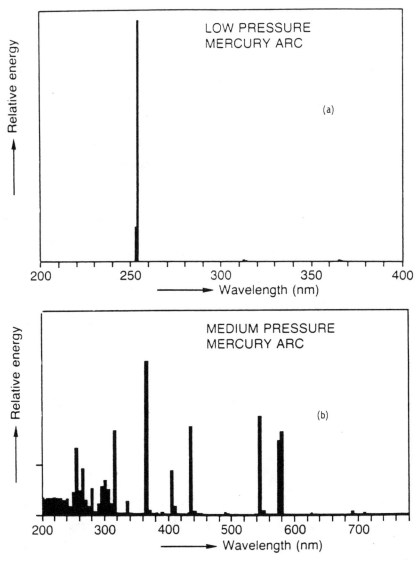

Figure 7.2   **Radiant power output spectra from (a) low-pressure and (b) medium-pressure mercury arc lamps (Meulemans, 1987)**

an expression of radiation energy per einstein allows direct comparison with bond energies per mole.)

Equation 7.1 can be used to show that radiation at $\lambda = 253.7$ nm has an associated energy of 112.8 kcal/einstein. A comparison of this value with the bond energies of several important bonds in microbial systems (see Table 7.1) reveals that radiation at 253.7 nm is sufficiently energetic to induce photochemical change.

Table 7.1 Bond energies of importance in microbiological systems (March, 1985)

Bond	Bond dissociation energy, kcal/mole
O-H	110–111
C-H	96–99
N-H	93
C=O	173–181
C-N	69–75
C=C	146–151
C-C	83–85

As described above, photochemical change is only possible if radiation energy is made available by absorption. Extensive research has shown that nucleic acids (such as deoxyribonucleic acid [DNA] and ribonucleic acid [RNA]) and proteins are effective absorbers of UV radiation (Jagger, 1967). In particular, these materials absorb strongly over the range $240 \leq \lambda \leq 260$ nm (see Figure 7.3). Because low-pressure mercury arc lamps emit the majority of their radiation at a wavelength within this range, they can be used effectively to induce a photobiochemical change in microorganisms.

Although proteins and nucleic acids are both effective absorbers of UV radiation, it is believed that the majority of UV-induced damage is imposed on

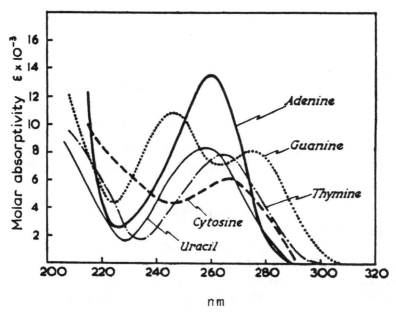

Figure 7.3 Ultraviolet absorption spectra for purine and pyrimidine bases at pH = 7 (Davidson, 1969)

the bases that compose nucleic acids. Dimerization of adjacent bases (especially thymine) on nucleic acid strands has been identified as the predominant UV inactivation mechanism (Jagger, 1967).

Alternative sources of UV radiation are also being investigated for disinfection processes. In particular, medium-pressure mercury arc lamps have been used for disinfection in some applications. The output spectrum of these lamps is substantially different from the spectrum of conventional low-pressure lamps (see Figure 7.2b). Radiation is emitted from these lamps over a large fraction of the UV spectrum. As a result, the responses of microorganisms to radiation from these lamps may be more complex than the responses elicited by exposure to radiation from low-pressure lamps. Furthermore, a theoretical analysis of photobiochemical change induced by medium-pressure lamps is more complex because of the polychromatic nature of the radiant energy source. However, the fundamental operation of disinfection processes that employ these lamps is conceptually similar to the operation observed for conventional low-pressure mercury arc lamps.

*P*HOTOREACTIVATION AND DARK REPAIR

Microorganisms have evolved and developed effective biochemical systems for repairing damage caused by hostile environmental conditions, such as exposure to disinfectants. Repair and recovery of sublethal damage is known to occur following all disinfection operations. Engineers involved in the design or operation of disinfection processes should understand these processes and their potential consequences.

Under some circumstances, the photobiochemical damage to an organism caused by UV irradiation can be repaired. These repair mechanisms allow UV-inactivated microorganisms to regain viability following the disinfection process. Two principal repair mechanisms have significance relative to UV disinfection: photoreactivation and dark repair.

Photoreactivation is a process whereby dimers within microbial nucleic acids are catalytically repaired to their original monomeric forms. Lindenauer and Darby (1994) summarized the current theory regarding the mechanism of photoreactivation. Reviews were also provided by Harm (1975) and the U.S. Environmental Protection Agency (U.S. EPA, 1986a). Observations of photoreactivation behavior can be explained using a two-step reaction mechanism (Figure 7.4). In the first step, a photoreactivating enzyme (PRE) combines with a pyrimidine dimer to form a PRE–dimer complex. The kinetics of this reversible reaction are such that the forward reaction (complex formation) is favored over the reverse reaction. Step 1 is a strict chemical reaction and, as such, requires no radiation to take place. In step 2, the PRE–dimer complex

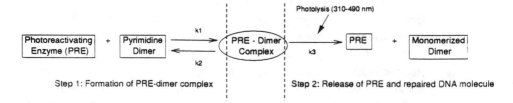

Figure 7.4   Hypothesized photoreactivation reaction mechanism (Harm, 1975, and Lindenauer and Darby, 1994)

absorbs radiation (310 nm $< \lambda <$ 490 nm), resulting in a photolytic reformation of the pyrimidine monomers and release of PRE. Reformation of the monomers results in reversal of photochemical damage. Once released from the PRE–dimer complex, PRE is available for further complex formation and photolytic repair.

The significance of photoreactivation will depend in large part on the initial dose of inactivating (UV) radiation, the dose of photoreactivating radiation, and the microorganism. As the UV dose increases, an exposed microorganism will accumulate more damage in the form of pyrimidine dimers. Therefore, reversal of that damage (sufficient to allow reactivation) will depend on the availability of photoreactivating radiation and PRE. Although most commonly associated with bacteria, photoreactivation has been observed in all taxonomic orders, including viruses (when in a host cell) (Jagger, 1967). However, some organisms do not demonstrate photorepair under field conditions (see Lindenauer and Darby [1994] for a review).

Dark repair provides a second mechanism for repairing photobiochemical damage. Dark repair processes are thought to involve enzymatic recognition of a dimer on a DNA strand. The dimer is excised from the DNA molecule, and the strand is repaired (U.S. EPA, 1986a). It is thought that dark repair processes have the ability to repair photoreactive and nonphotoreactive damage (Jagger, 1967).

The existence of repair mechanisms for UV-induced damage opens several ongoing questions, the primary one being to what extent these mechanisms should be taken into consideration during design. From an operational perspective, the availability of repair mechanisms would dictate a larger UV dose than would be required if no repair were possible. Similarly, the inclusion of reactivation mechanisms in the design process requires more UV hardware. Lindenauer and Darby (1994) suggest in their analysis that the effect of photoreactivation is relatively insignificant at the dose levels they interpreted (that is, reuse applications at doses greater than 60 to 80 mW·s/cm^2). However, much of the design work for secondary WWTPs has been at equivalent dose levels of less than 40 mW·s/cm², at which point significant increases in residual densities have been measured (via the static light/dark bottle technique) (U.S. EPA, 1986a).

Lehrer and Cabelli (1993) pointed out that the etiologic agents of the most common waterborne diseases are Norwalk-like viruses. Many of these viruses are not thought to undergo repair of UV-induced damage. Although quantification of the Norwalk virus is quite difficult, inactivation of physiologically similar viruses by UV disinfection has been shown to be quite effective, especially when compared with chlorine-based disinfection processes (Yip and Konasewich, 1972).

Whitby and Palmateer (1993) suggest that this phenomenon is not observed *in situ*. Using labeled *Escherichia coli* (*E. coli*) bacteria, they demonstrated a lack of reactivation in UV-irradiated wastewater effluent after release to a receiving stream. These same bacteria were shown to undergo photoreactivation when exposed to a sufficient dose of photoreactivating radiation under controlled conditions. One possible explanation for this discrepancy is that organisms released to receiving water did not receive a sufficient dose of photoreactivating radiation to undergo repair. The dose of photoreactivating radiation received by organisms in wastewater effluent will be site-specific and will depend on factors such as water quality, receiving water depth, and dilution in receiving water. While the study of Whitby and Palmateer (1993) demonstrates that photoreactivation may not always be significant, designers of UV systems should understand that reactivation can occur under certain environmental conditions.

One design approach that has been taken is to isolate the effect of photorepair by using sampling techniques that allow measurements of coliform that have not undergone photorepair, then measuring the extent to which repair can take place (maximum photoreactivation) by exposing them to visible light in a transparent bottle. The design sizing of the system is estimated on the basis of no repair and then increased to accommodate some level of repair. In effect, this means that the equivalent dose is increased by a factor. Although this is a conservative design approach, it does offer protection against some practical issues:

- Sampling at most WWTPs is done at a location downstream of the UV disinfection system, such that some degree of repair can occur in the whole effluent (for example, postaeration). Repair can also occur in the containers themselves if they are transparent to visible light. Thus, monitoring at an operating WWTP may in itself induce the effect of repair to a greater degree than may occur *in situ*. Some regulatory agencies actually require this (for example, allowing for repair to occur in the sample before the densities are determined analytically).
- If certain pathogens have the ability to repair, their effect should be considered in design as a matter of protocol. In effect, the reuse work

in California has pointed to this by the imposition of high dose levels. When the procedure suggested by U.S. EPA (1986a) is used, the dose levels are increased by a factor of two to three to account for photoreactivation, yielding equivalent doses on the order of 60 to 100 mW·s/cm^2.

A consensus does not exist within the engineering or regulatory communities regarding the inclusion of repair in UV disinfection system design. Although repair and recovery mechanisms are known to exist for other disinfectants, including chlorine (see Calmer *et al.*, 1994), little attention is generally given to the subject. Although many operating WWTPs have been designed and are operating successfully with and without consideration of repair, one should be cautioned that this does not mean photorepair is not occurring, but that it may be masked by overdesign, underuse (WWTPs are well below their design capacity), and sampling/analysis techniques. Overall, the reader is cautioned about the effects of repair and should be cognizant of the potential impacts of designing a system with or without its consideration.

*U*LTRAVIOLET INACTIVATION KINETICS

In evaluating the rate of photobiochemical change induced by UV irradiation, it may be useful to view a UV photon as a "reactant." As in the case of strict chemical reactions, the rate of a photochemical reaction is governed by the availability of reactants. In a chemical reaction, reactant availability is quantified using concentration or activity. Intensity is the measure of radiation "availability" in a photochemical reaction. Therefore, knowledge of radiation intensity is required for estimating photochemical kinetics.

The experimental procedure used to evaluate inactivation dose-response behavior (kinetics) involves exposure of a microbial population to a measurable source of radiation for a known period of time, followed by quantification of microbial viability. In most cases, the source of radiation is a collimated beam. An example of a collimation apparatus is illustrated in Figure 7.5. The purpose of the collimator is to produce radiation that is nearly parallel and can be imposed perpendicular to a planar surface. This is necessary because the radiometers used in these experiments are designed to quantify radiation intensity that is perpendicular to the detector surface.

Microbial irradiation is accomplished by placement of a shallow petri dish in the collimated beam. Water depth is typically maintained at 10 mm or less, and the liquid containing the microbial population is kept well mixed by the use of a micromagnetic stir bar. The purpose of mixing is to ensure uniform irradiation of all microorganisms in the liquid.

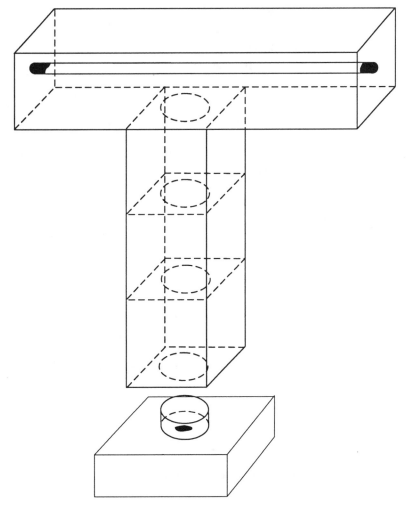

**Figure 7.5   Schematic illustration of collimation apparatus for
ultraviolet exposure experiments**

Because radiation energy may be absorbed by the medium in which the
microorganisms are suspended, the actual intensity within the medium will
decrease from its top surface downward. Given this nonuniform intensity
field, it may be necessary to calculate depth-averaged intensity within the
entire irradiated volume. The variation in intensity with depth may be calculated by applying Beer's law:

$$I_x = I_o e^{-\alpha x} \tag{7.2}$$

Where

I_x = radiation intensity at depth x ($x \geq 0$), mW/cm^2;
I_o = incident radiation intensity, mW/cm^2;
α = absorbance coefficient, cm^{-1}; and
x = depth, cm.

(Note: A more complete description of Beer's law is given in the section titled Intensity.)

This equation may be integrated over the entire fluid depth ($0 \leq x \leq H$) to yield a depth-averaged intensity for the entire reactor:

$$I_{avg} = \frac{I_o}{H\alpha} \left(1 - e^{-H\alpha}\right) \qquad (7.3)$$

Where

H = liquid depth, cm.

If the liquid within the reactor is kept shallow ($H \leq 1$ cm) and well-mixed, I_{avg} can be used to characterize the entire medium with little error.

As a first approximation, inactivation kinetics in a well-mixed, shallow petri dish can be modeled as a first-order photobiochemical reaction,

$$\frac{dN}{dt} = -kI_{avg}N \qquad (7.4)$$

Where

N = concentration of viable organisms, no. organisms/L;
t = time, s;
I_{avg} = average radiation intensity, mW/cm^2; and
k = first-order inactivation constant, cm^2/mW·s.

By integration,

$$\frac{N}{N_o} = e^{-kIt} \qquad (7.5)$$

Where

N_o = concentration of viable organisms before exposure, no. organisms/L; and
It = ultraviolet "dose," mW·s/cm^2.

The results of collimated beam experiments have been reported for many microorganisms. A summary of recently reported dose-response behaviors is presented in Table 7.2. It should be noted that the inactivation data in Table 7.2 were compiled from a large number of independent experiments conducted by different researchers and using different techniques. Therefore, direct, literal comparison of results from these experiments may, at times, be misleading. Nonetheless, these data reveal several interesting characteristics of microbial inactivation by UV irradiation. Generally speaking, viruses and bacteria are inactivated effectively by UV irradiation, whereas protozoan cysts and spore-forming bacteria are relatively resistant to inactivation.

Most wastewater disinfection processes are regulated by effluent coliform viability. The coliform bacteria are used as indicator organisms to assess microbial activity in effluents. Historically, coliform bacteria have been used for this purpose because it has been found that in chlorine-based disinfection processes, the conditions required to inactivate coliform bacteria adequately will also achieve sufficient inactivation of other microorganisms and prevent communicable disease transmission. The data presented in Table 7.2, as well as other information (Yip and Konasewich, 1972), indicate that the UV dose required to meet coliform limitations will achieve better virus inactivation than the comparable chlorine dose. Viruses are suspected to be the etiologic agents responsible for many waterborne disease outbreaks (Craun, 1988).

The first-order model represents a reasonable first approximation to UV dose-response behavior. For small organisms (such as viruses), first-order behavior has been observed over as much as 5 logs of inactivation (Harris *et al.*, 1987). First-order behavior is also commonly observed for lesser inactivations in bacteria (that is, 3-log inactivation or less). However, two important deviations from first-order behavior have been observed (see Figure 7.6). First, a lag in microbial activation is sometimes noted at low doses; second, a decline in the slope of the dose-response curve is seen at high doses (tailing). Several hypotheses have attempted to describe this behavior.

The lag may be attributable to the ability of microorganisms to absorb a sublethal dose of radiation without showing any adverse effects in the analytical procedures used to quantify their viability. The existence of such a "threshold" has been noted in other disinfection processes as well.

Tailing may be the result of heterogeneity among a population of microorganisms. Some organisms may be weak or otherwise relatively susceptible to inactivation by UV exposure, whereas other organisms in the same population may be resistant to exposure. Another possible cause of tailing may be the presence of particles. Particles may shade microorganisms simply by providing an opaque surface to incident UV radiation, or they may shield an organism from UV radiation by incorporating a viable organism within the particle matrix.

The effects of particles on bacterial inactivation by UV irradiation have been examined by many researchers. Oliver and Cosgrove (1975) observed

---

Table 7.2   Reported microbial dose-response behavior resulting from ultraviolet irradiation

Group	Microorganism	Dose required for 90% inactivation, mW·s/cm^2	First-order inactivation constant, cm^2/mW·s	Reference
Bacteria	*Aeromonas hydrophila*	1.54	1.50	1[a]
	Bacillus anthracis	4.5	0.51	2[b]
	Bacillus anthracis spores	54.5	0.042 2	2
	Bacillus subtilus spores	12	0.19	2
	Campylobacter jejuni	1.05	2.19	1
	Clostridium tetani	12	0.19	2
	Corynebacterium diptheriae	3.4	0.68	2
	Escherichia coli	1.33	1.73	1
	Escherichia coli	3.2	0.72	2
	Escherichia coli	3	0.77	3[c]
	Klebsiella terrigena	2.61	0.882	1
	Legionella pneumophila	2.49	0.925	1
	Legionella pneumophila	1	2.3	2
	Legionella pneumophila	0.38	6.1	3
	Micrococcus radiodurans	20.5	0.112	2
	Mycobacterium tuberculosis	6	0.38	2
	Pseudomonas aeruginosa	5.5	0.42	3
	Pseudomonas aeruginosa	5.5	0.42	2
	Salmonella enteris	4	0.58	3
	Salmonella enteritidis	4	0.58	2
	Salmonella paratyphi	3.2	0.72	2
	Salmonella typhi	2.26	1.02	1
	Salmonella typhi	2.1	1.1	2
	Salmonella typhi	2.5	0.92	3
	Salmonella typhimurium	8	0.29	2
	Shigella dysentariae	2.2	1.05	2
	Shigella dysentariae	0.885	2.60	1
	Shigella dysentariae	2.2	1.05	3
	Shigella flexneri	1.7	1.4	3
	Shigella paradysenteriae	1.7	1.4	3
	Shigella sonnei	3	0.77	3
	Staphylococcus aureus	5	0.46	2

dysentry

Table 7.2 **Reported microbial dose-response behavior resulting from ultraviolet irradiation (continued)**

Group	Microorganism	Dose required for 90% inactivation, mW·s/cm^2	First-order inactivation constant, cm^2/mW·s	Reference
	Staphylococcus aureus	4.5	0.51	3
	Streptococcus faecalis	4.4	0.52	2
	Streptococcus pyogenes	2.2	1.0	2
	Vibrio cholerae	0.651	3.54	1
	Vibrio cholerae - cholera	3.4	0.68	3
	Vibrio comma	6.5	0.35	2
	Yersinia enterocolitica	1.07	2.15	1
Viruses	Coliphage	3.6	0.64	3
	Coliphage MS-2	18.6	0.124	1
	F-specific bacteriophage	6.9	0.33	2
	Hepatitis A	7.3	0.32	1
	Hepatitus A	3.7	0.62	3
	Influenza virus	3.6	0.64	2
	Poliovirus	7.5	0.31	2
	Poliovirus 1	5	0.5	3
	Poliovirus type 1	7.7	0.30	1
	Rotavirus	11.3	0.204	2
	Rotavirus SA-11	9.86	0.234	1
	Rotavirus SA-11	8	0.3	3
Protozoa	Giardia muris	82	0.028	3
	Acanthamaoeba castellanii	35	0.066	3

[a] 1 = Wilson *et al.* (1992).
[b] 2 = Cairns (1991).
[c] 3 = Wolfe (1990).

significant reductions in inactivation of heterotrophic bacteria in raw wastewater as compared with secondary effluent at comparable UV doses. They attributed the decrease in bacterial kill to the presence of suspended particles. After exposing the fluid to ultrasonication, they observed an improvement in kill resulting from the UV exposure. Although no data were provided, the improved kill was attributed to the breakup of large aggregated particles.

Qualls *et al.* (1983) developed dose-response curves for coliform bacteria in unfiltered secondary effluent. Dose-response curves were also developed on separate subsamples of the same secondary effluent after being passed through 8-μm or 70-μm filters. The dose-response curves for these waters are illustrated in Figure 7.7. Because coliform bacteria are 1 to 2 μm in size, it is

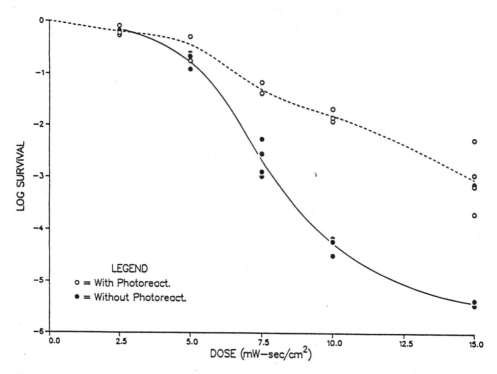

Figure 7.6   Observed deviations from ideal first-order dose-response behavior resulting from ultraviolet irradiation (Harris *et al.*, 1987)

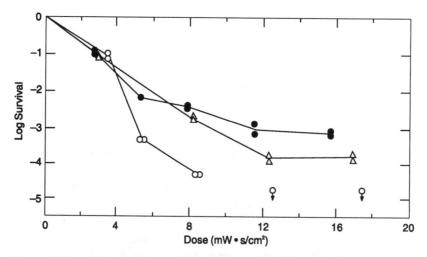

Figure 7.7   Effect of filtration on coliform survival in wastewater effluent; arrows indicate limit of detection (○ = filtered through 8-μm filter; Δ = filtered through 70-μm filter; ● = unfiltered) (Qualls *et al.*, 1983)

likely that only single bacteria or small bacterial clumps were allowed to pass through the 8-μm filter. The 70-μm filter could allow passage of some comparatively large particles. Filtration through the 70-μm filter allowed minimal improvement in kill as compared with the unfiltered sample; passage through the 8-μm filter allowed a greater improvement in kill. The conclusion from this work was that removal of relatively large ($d_p \approx 70$ μm) particles allowed substantial improvement in disinfection because these particles could harbor many bacteria and effectively shade them from UV exposure.

Darby *et al.* (1993) examined coliform kill in sand-filtered and unfiltered secondary effluent. Analysis of the particle size distribution (PSD) in both waters revealed a bimodal distribution with peaks near particle sizes of 1 μm and 35 μm. Filtration achieved removal of 40% of the particle volume for particles in the 0.6 to 1.3 μm diameter range and 64% removal for particles in the 15.8 to 63 μm diameter range (see Figure 7.8). Coliform kills were consistently higher in the filtered effluent as compared with the unfiltered effluent. The improvement probably resulted from the removal of the larger particles (15.8 to 63 μm diameter), which otherwise could harbor (shade) bacteria from UV exposure.

It should be noted that most investigations of particles and their effects on microbial inactivation have focused on coliform bacteria. The conclusion of these investigations has been that bacteria (typical size $\approx$ 1 μm) can be protected from UV exposure by large particles. It is likely that other microorganisms of interest (such as viruses: typical size $\approx$ 0.01 μm) will display

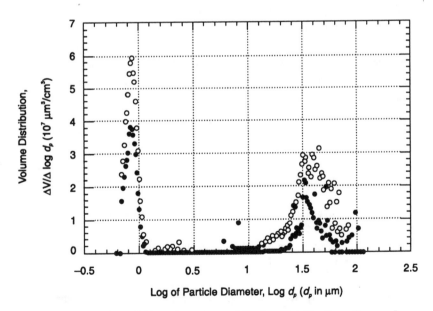

Figure 7.8   **Effect of filtration on particle size distribution of secondary effluent from a wastewater treatment facility (○ = unfiltered; • = filtered) (Darby *et al.*, 1993)**

---

qualitatively similar behavior. However, the critical particle size for protection of viruses from UV irradiation is unknown.

OTHER KINETIC MODELS

A number of models have been developed to account for the nonideal factors in UV dose-response behavior. A few of these models will be described briefly to provide the reader with background on the subject. Additional detail may be gained from the references themselves.

Severin *et al.* (1983 and 1984) presented two models that account for lag behavior in UV disinfection. The multitarget model was developed on the assumption that an individual organism or clump of organisms will contain a finite number (n_c) of critical targets; all critical targets must be "hit" by a photon of UV radiation for the organism (or clump) to appear inactivated in a subsequent analytical procedure. The probability of any critical target surviving irradiation is assumed to be governed by a first-order relationship:

$$P(0) = e^{-kIt} \tag{7.6}$$

Where

$P(0)$	=	probability of target survival;
k	=	rate constant, $cm^2/mW \cdot s$;
I	=	radiation intensity; mW/cm^2; and
t	=	period of exposure, s.

If all targets are assumed to be randomly distributed and photochemically equivalent, then the fractional survival of an irradiated population may be described by the following:

$$\frac{N_s}{N_I} = 1 - (1 - e^{-kIt})^{n_c} \tag{7.7}$$

Where

N_s	=	concentration of viable organisms in irradiated sample, no. organisms/L; and
N_I	=	concentration of viable organisms in unirradiated sample, no. organisms/L;

Note that the multihit model reduces to a simple first-order model when $n_c = 1$.

The values of the parameters k and n_c will be specific to each organism and a given set of environmental conditions. Parameter estimation can be accomplished through a least-squares fit of batch inactivation data to Equation 7.7.

The series-event model was developed under the assumption that inactivation of organism (or particle) elements takes place in a serial manner. An organism will remain viable until a threshold has been reached, whereby n elements of the organism have been inactivated. Each "event" is assumed to represent a discrete unit of photobiochemical damage within the organism. The rate of event occurrence is assumed to be identical for all levels and independent of the event level achieved by an organism. The photobiochemical reaction sequence can be represented by the following:

$$M_o \xrightarrow{kI} M_1 \xrightarrow{kI} \ldots M_i \xrightarrow{kI} \ldots M_{n-1} \xrightarrow{kI} M_n \xrightarrow{kI} \ldots \qquad (7.8)$$

Where

M_i = an organism that has reached level i, and

n = threshold level of organism.

Individual event kinetics are assumed to be first-order, such that

$$r_{N_i} = \frac{dN_i}{dt} = kIN_{i-1} - kIN_i \qquad (7.9)$$

Where

r_{N_i} = rate at which organisms pass through event level i, no. organisms/L·s.

An organism will be inactivated if UV exposure brings it to event level n or higher. Therefore, all surviving organisms will have reached level $n - 1$ or less. Using this logic, it can be shown that the fractional survival among exposed organisms is described by

$$\frac{N_s}{N_I} = e^{-kIt} \sum_{i=0}^{n-1} \frac{(kIt)^i}{i!} \qquad (7.10)$$

The series-event model also reduces to simple first-order kinetics when $n = 1$. The parameters k and n will be specific to a given organism and set of conditions. Estimates of model parameters can be derived using the procedure described above for the multitarget model.

Severin *et al.* (1983) estimated model parameters (k and n_c for the multihit model; k and n for the series-event model) for pure cultures of *Candida parapsilosis*, *E. coli*, and coliphage virus f2 with host bacteria *E. coli* K-13 (see Table 7.3). Although both models could be used to fit experimental data effectively ($r^2 > 0.95$, in all cases; see Figures 7.9 and 7.10), the series-event model was regarded as being superior because it was assumed to provide a more accurate representation of the actual mechanism of photobiochemical inactivation in UV systems.

Table 7.3   Multihit and series-event model parameters for inactivation kinetics (Severin *et al.*, 1983)

Organism	Multitarget model		Series-event model	
	n_c	k (cm^2/mW·s)	n	k (cm^2/mW·s)
Escherichia coli	201	0.893	9	1.538
Candida parapsilosis	1 871	0.433	15	0.891
Coliphage f2	1	0.072 4	1	0.072 4

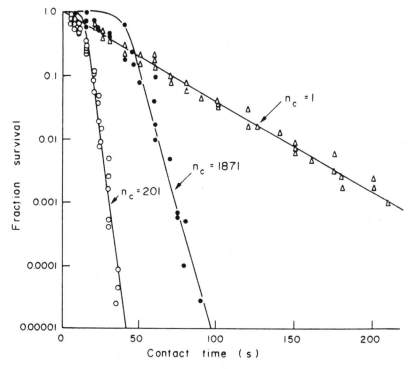

Figure 7.9   Application of multitarget kinetic model to inactivation response of *E. coli* (○), *C. parapsilosis* (•), and coliphage f2 (Δ) in batch reactors (from Severin, B.F., *et al.* [1983] Kinetic Modeling of U.V. Disinfection of Water. *Water Res.*, 17, 1669)

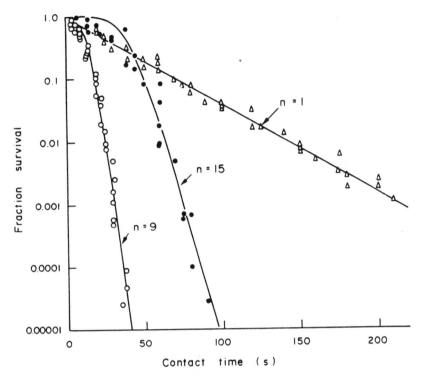

Figure 7.10   **Application of series-event kinetic model to inactivation response of *E. coli* (○), *C. parapsilosis* (•), and coliphage f2 (Δ) in batch reactors (from Severin, B.F., *et al.* [1983] Kinetic Modeling of U.V. Disinfection of Water. *Water Res.*, 17, 1669)**

Collins and Selleck (1972) developed a model that could be used to describe lag and tailing behavior. As with most other disinfection models, the Collins-Selleck model is a modification of conventional first-order kinetics. In their model, the fundamental equation used to describe microbial inactivation is

$$\frac{dN}{d(It)} = -kN \qquad (7.11)$$

Where

k = reaction parameter, cm^2/mW·s;
It = UV dose, mW·s/cm^2; and
N = viable organism concentration, no. organisms/L.

The product *It* is taken as the independent variable; the reaction parameter is allowed to vary with time (dose) as follows:

$$k = 0 \text{ for } It \le D$$

$$k = \frac{K_0}{It} \text{ for } It > D \qquad (7.12)$$

Where

K_0 = constant; and

D = threshold dose required to achieve inactivation, mW·s/cm^2.

By defining a boundary condition of

$$N = N_o \text{ for } It \le D \qquad (7.13)$$

the Collins-Selleck model yields

$$\frac{N}{N_o} = 1 \text{ for } It \le D$$

$$= \frac{It}{D}^{-K_0} \text{ for } It \le D \qquad (7.14)$$

The semiempirical approach used in developing the Collins-Selleck model allows lag and tailing behavior to be predicted. The model was originally used to predict the effect of chlorination on the inactivation behavior of coliform bacteria in secondary effluents, but it can be also applied to alternative disinfectant systems in which the same phenomena (lag and tailing) are observed.

EFFECT OF INTENSITY ON INACTIVATION BEHAVIOR

Studies have been performed using UV lasers to evaluate bacterial inactivation at high radiation intensities. Oliver and Cosgrove (1975) used an argon ion laser ($\lambda = 257.3$ nm) and a series of filters to vary UV intensity applied to a flow-through, bench-scale UV disinfection system. The inactivation responses of heterotrophic bacteria were essentially identical over the range of applied intensities for these experiments (1 mW/cm^2 < I < 200 mW/cm^2). Zubrilin et al. (1991) reached similar conclusions using a krypton monofluoride laser ($\lambda = 248$ nm) as the source of radiation. In both studies, bacterial inactivation was observed to be a function of dose but not of applied intensity. Hunt (1992) developed dose-response curves for *Bacillus subtilis* spores using collimated UV radiation at intensities ranging from 0.070 to 1.0 mW/cm^2. Dose-response behavior was similar for these organisms at all intensities. The results of these experiments suggest that antimicrobial efficacy resulting from

UV irradiation is a function of the total "dose" of radiation received but is independent of the "path" used to deliver that dose.

In full-scale, open-channel UV systems, microorganisms are exposed to a rapidly varying intensity field within the UV lamp array. Typical UV intensities within these systems are in the range of 1 to 20 mW/cm^2. Based on the experimental results described above, it would appear that bacterial inactivation is independent of intensity for the range of intensities observed in the field. The UV dose will dictate the amount of microbial damage, independent of the path used to apply the dose.

In continuous-flow UV disinfection systems, microorganisms will be exposed to conditions of variable intensity. Therefore, a more accurate description of the UV dose might be

$$\text{Dose} = \int_0^{\tau} I(t) \; dt \qquad (7.15)$$

Where

$I(t)$ = time-dependent intensity function, mW/cm^2; and

τ = period of exposure, s.

At present, this analysis remains strictly theoretical; available information does not allow prediction of $I(t)$. However, the concept of a time-integrated dose is useful in a discussion of intensity and hydraulics.

The preceding discussion illustrates the response of microbial pathogens and indicator organisms to UV irradiation. In flow-through systems, the goal is to deliver a sufficient dose to the microbial population to ensure adequate inactivation. As in collimated beam experiments, the delivered dose in a flow-through UV system will depend on the intensity of radiation and duration of exposure. Therefore, an analysis of flow-through systems requires an understanding of the intensity distribution within a UV lamp array and of the hydrodynamic behavior within the system.

*I*NTENSITY

The UV irradiation intensity is a measure of radiative power per unit of exposed area. The distribution of intensity within a UV lamp array is highly nonuniform but can be predicted with reasonable accuracy using straightforward numerical techniques. The geometry of a mercury arc lamp does not lend itself to strict numerical analysis for predicting intensity distributions. Therefore, lamp geometry is artificially simplified by assuming that a lamp may be represented by a series of colinear point sources located along the axis of the lamp. The total output power of this series of point sources is defined as

---

being equal to the output power of the actual lamp. Intensity at any given receptor site is then defined as the sum of intensity contributions from all point sources in the system.

The variation in intensity from a single point source is assumed to be attributable to two phenomena: dissipation and absorbance. In describing the dissipation mechanism, it is useful to envision a single point source with known radiative power emitting radiation in all directions (see Figure 7.11). Assume for now that no other mechanism of intensity variation exists in the system (that is, no absorbance). Because radiation is emitted uniformly in a radial pattern, the power received by any spherical surface with an origin that coincides with the point source will be identical. Using similar logic, it may be argued that the intensity of radiation received by such a spherical surface will depend on the area over which that power is distributed, namely, the area of a sphere of radius ρ, where ρ = (radial) distance from point source to receptor location. Therefore, the spatial variation in intensity resulting from dissipation may be defined as follows:

$$I(\rho) = \frac{P/n}{4\pi\rho^2} \qquad (7.16)$$

Where

$I(\rho)$ = intensity at distance ρ from the point source, mW/cm^2;

ρ = radial distance from the point source to the receptor location, cm;

P = lamp output power, mW;

n = number of point sources being used to model lamp; and

P/n = power per point source, mW.

As with any form of electromagnetic radiation, UV intensity will also vary by the mechanism of absorbance. Beer's law states that the gradient in intensity is linearly related to the intensity itself:

$$\frac{dI}{dx} = -\alpha \cdot I \qquad \text{or} \qquad I(l) = I_o e^{-\alpha l} \qquad (7.17)$$

Where

x = distance in the direction of irradiation, cm;

α = absorbance coefficient, cm^{-1};

I_o = intensity of incident radiation, mW/cm^2; and

l = path length in absorbing medium, cm.

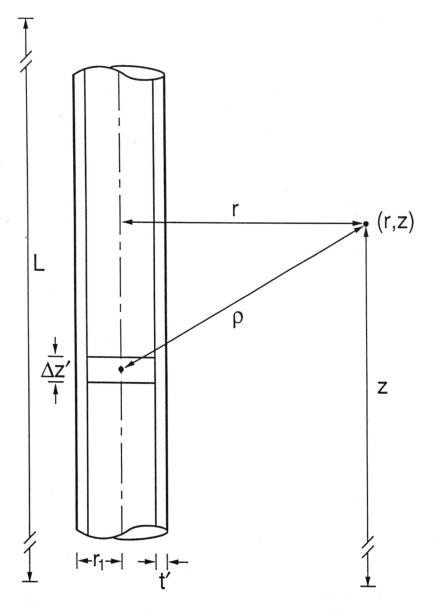

Figure 7.11   Schematic illustration of lamp geometry as used in point-source summation technique (Jacob and Dranoff, 1970)

Ultraviolet radiation at $\lambda = 253.7$ nm will be absorbed by the quartz jackets that surround the lamps and the water being irradiated. The absorptive behavior of both media may be accounted for using an expression analogous to Equation 7.17 above. Total absorbance in the system can be quantified as follows:

$$I = I_o \exp\left[-(\alpha_q l_q + \alpha_l l_l)\right] \qquad (7.18)$$

Where

α_q = absorbance coefficient for quartz, cm^{-1};
α_l = absorbance coefficient for liquid, cm^{-1};
l_q = path length for absorbance in quartz, cm; and
l_l = path length for absorbance in liquid, cm.

The terms l_q and l_l may be defined more explicitly in terms of system parameters through the application of geometry (Jacob and Dranoff, 1970) (see Figure 7.11). Using this approach and combining Equations 7.16 and 7.18 above, the following equation may be used to define the intensity contribution at any receptor site from point source i:

$$I(r, z)_i = \frac{P/n}{4\pi\rho^2} \exp\left\{-[\alpha_q \, t' + \alpha_l \, (r - r_1)]\frac{\rho}{r}\right\} \qquad (7.19)$$

Where

ρ = $[r^2 + (z - h)^2]^{1/2}$, cm;
t' = thickness of quartz sleeve, cm;
z = longitudinal coordinate of receptor site, cm;
h = longitudinal coordinate of point source, cm; and
r = radial distance from lamp axis to receptor site, cm.

The total intensity received at any location is then estimated as the sum of intensity contributions from all point sources in the system:

$$I(r, z) = \sum_{i=i}^{n} I(r, z)_i \qquad (7.20)$$

Variations of Equations 7.19 and 7.20 have been used extensively to predict intensity distributions within UV arrays. The spatial distribution of UV intensity within a lamp array displays variation in all three dimensions and is a strong function of the absorptive behavior within the system. The average intensity within the reactor is computed by averaging the volumes of single-point intensity estimates. Beyond this, adjustments must be made to account for aging of the lamp and fouling of the quartz surfaces through which the energy is being transmitted.

The simplest system for application of point-source summation would involve a single lamp. A plot of intensity around a single lamp at three radial distances from the lamp jacket surface $(r - r_1)$ is presented in Figure 7.12.

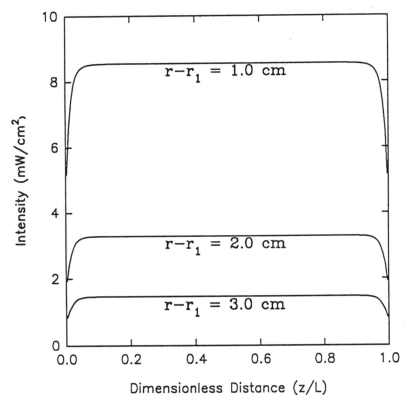

Figure 7.12 Intensity distributions calculated by point-source summation for 147-cm (1 470-mm) arc length low-pressure mercury arc lamp; the dimension $r - r_1$ represents radial distance from the outside of the quartz jacket to the receptor site

Intensity is observed to decrease rapidly with radial distance from the lamp. Moreover, an area of intensity depression is evident in the vicinity of the lamp electrodes ($z/L \approx 0$; $z/L \approx 1.0$). Figure 7.13 presents the results of a point-source summation analysis with several lamps of different lengths. If the "end-effects region" is arbitrarily defined as the region in which $I(r,z) \leq 95\%$ of the peak intensity (for a fixed value of $r - r_1$), then the size of the end-effects region can be calculated. It can be seen from this figure that the size of the end-effects region is a function of the size of the lamp used (Blatchley *et al.*, 1995). In relative terms, long arc-length lamps have small end-effects regions; it is therefore advantageous to employ long lamps whenever possible.

The intensity field within an array of lamps may be calculated similarly by adding the contributions of all lamps in the system. Figure 7.14 presents the results of such an analysis for three different conditions of transmittance ($T =$ 40%, $T = 65\%$, and $T = 90\%$; path length = 1.0 cm; and $\lambda = 253.7$ nm). The effect of transmittance is evidenced by the following predicted intensity

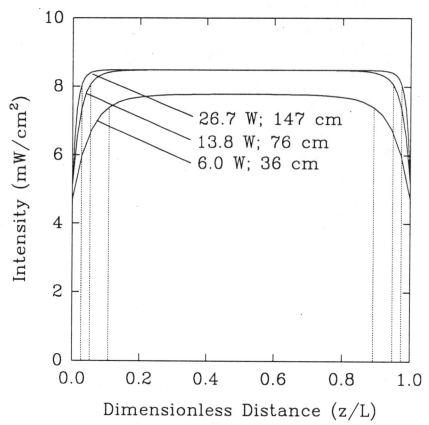

Figure 7.13 Intensity distributions at a radial distance of 1.0 cm (10 mm) from the outside of quartz lamp jackets for low-pressure mercury arc lamps used in open-channel ultraviolet disinfection systems; end effects regions for each lamp are indicated by dotted vertical lines (Blatchley *et al.,* 1995)

distributions: a low value of T will yield an intensity distribution that contains large areas of low intensity; a high value of T allows high intensity values, even at relatively large distances from the lamps within the array.

Areas of low intensity represent potential problems in UV systems because the UV dose received by an organism is a function of the intensity of radiation delivered. Therefore, a low value of transmittance may render the application of UV disinfection difficult or impossible. A value of 65% T ($\lambda = 253.7$ nm, and path length = 1.0 cm) is typically used as a rule-of-thumb lower limit for determining the potential for success in conventional open-channel UV systems. It should be noted that UV disinfection has been applied successfully to waters in which T is less than or equal to 65%. Furthermore, "nonconventional" systems (such as systems with tighter lamp spacing or high-intensity lamps) may be used with waters that display poor transmittance characteristics.

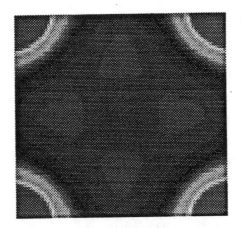

T = 40%

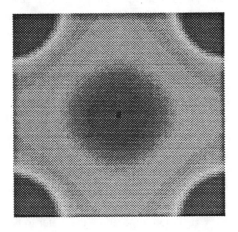

T = 65%

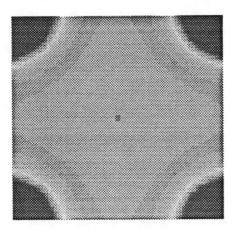

T = 90%

Figure 7.14   Intensity fields calculated by point-source summation
within an array of four lamps with axes at the corners of a
7.6 cm × 7.6 cm square; fields are displayed for water
transmittances of 40, 65, and 90%

Even in systems with high T and long lamps, strong intensity gradients will exist within a UV lamp array. As the most distant point from the lamp jacket surfaces, the areas of lowest intensity also coincide with areas of highest average velocity. Therefore, the potential exists for a flow pattern that would allow some microorganisms to experience both low intensity and short exposure, while other organisms experience high intensity and long exposure. This pattern will yield inefficient use of UV radiation from the standpoint of microbial kill.

The effect of the scenario described above can be minimized if fluid flow within the system allows mixing across intensity gradients. Qualitatively speaking, mixing will be promoted by turbulent flow conditions. Therefore, UV systems should be operated at a sufficiently high approach velocity to ensure turbulent conditions. It should be noted that while turbulent flow conditions will promote mixing, these conditions do not guarantee a condition of complete mixing in any direction within a UV lamp array. Severin *et al.* (1983 and 1984) demonstrated that extremely aggressive mechanical agitation was required to achieve a condition of complete mixing in a small, bench-scale laboratory system. It is highly unlikely that the flow regimen within an open-channel UV system will allow sufficient disturbance to achieve a condition of complete mixing. Therefore, some degree of stratification will exist within the irradiated zones of open-channel UV systems, diminishing the efficiency of UV radiation for disinfection.

Designers of UV systems should note that the UV intensity field for a system will not be constant. As described above, changes in water quality (in other words, UV transmittance) will alter the intensity field. Of similar importance are the effects of lamp age and lamp fouling. These two factors combine to reduce the amount of UV radiation imposed on the water to be disinfected. A more detailed description of these effects can be found in the Fouling section.

The point-source summation (PSS) procedure can be used to estimate average intensity for a given set of conditions. Conceptually, this procedure involves calculation of the intensity field within an array by PSS. These data are integrated over the volume of the irradiated zone to provide an estimate of the average intensity within the array.

The computer program UVDIS 3.1 (HydroQual, 1992b) was developed for design and analysis of UV systems. The subroutine "TULIP" within this program allows calculations of array-averaged intensity using PSS. Although it has been widely used, the program itself requires further work and refinement, particularly when considering advanced lamp systems. UVDIS 3.1 operates on most personal computers and has been recommended as a standard for average intensity estimation in UV systems (NWRI, 1993). Its algorithms use the PSS method but incorporate several assumptions and factors internally to simplify the calculation and shorten the computing time. These factors include transmission losses through the lamp envelope and quartz sleeve and the spacing and frequency of calculation iterations.

ULTRAVIOLET DOSE

Data from PSS analyses are often used to estimate the UV dose to be delivered by a continuous-flow system as follows:

$$\text{Dose} = I_{avg} \cdot \theta \qquad (7.21)$$

Where

Dose	=	average UV dose delivered, mW·s/cm^2;
I_{avg}	=	array-averaged intensity from PSS, mW/cm^2; and
θ	=	mean hydraulic detention time within irradiated zone, s.

The value of θ may be estimated using knowledge of the hydraulic loading and system geometry. Alternatively, θ may be estimated using tracer techniques (see discussion in Hydraulics section). A considerable amount of data exists on dose estimation using this approach.

The bioassay method has also been used for dose estimation in UV systems. This procedure is illustrated schematically in Figure 7.15. The dose-response behavior of an indicator organism is first quantified (Figure 7.15, right) using a collimated UV source for batch irradiation (Figure 7.15, left). Typically, a *Bacillus subtilis* spore has been used, but other organisms, including indigenous bacteria, may also be employed for this purpose (Blatchley and Hunt, 1994).

The calibrated organism is then injected into the influent stream of the test system on a continuous basis. Once a steady state is achieved, samples are taken of the influent and effluent to determine the response of the organism.

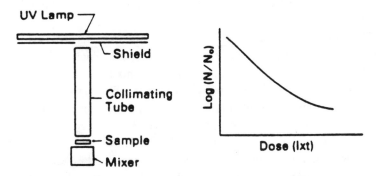

Figure 7.15   Schematic representation of bioassay procedure (left, dosing apparatus; right, dose-response calibration) (U.S. EPA, 1986a)

From this, a delivered dose can be implied using the dose-response calibration developed earlier. The procedure is repeated over a range of flow conditions, leading to a dose–flow relationship for the specific piece of UV equipment. This procedure is useful for comparing different system configurations and examining the effects of specific modifications (such as spacing, hydraulics, and lamp output). However, one should exercise caution, understand the inherent variability of the test and implications of the test results, and ensure that the protocols used conform to general practice.

Each method of dose estimation has advantages and disadvantages. The PSS method is relatively inexpensive and simple to apply, and can be compared with dose estimates from many other sites. The principal advantage of the bioassay is that the test is performed with microorganisms and provides a measure of actual microbial response.

A disadvantage of both techniques is that a single number ("dose") is used to characterize the system when, in fact, real UV systems will deliver a varying distribution of UV doses. Furthermore, the physical meanings of these two doses are fundamentally different. As a result, dose estimates from the two procedures can be quite different, even when the procedures are applied to the same system at the same time.

The differences in dose estimates using PSS and bioassay are attributable to the assumptions used in each procedure. A detailed discussion of these assumptions and their physical significance in UV systems was presented by Blatchley and Hunt (1994). The PSS procedure uses a single value of intensity (I_{avg}) and a single period of exposure (θ) to characterize the dose delivered by the system. In real UV systems, the intensity of exposure for any fluid element will be a rapidly varying function that will depend on the physical location of the fluid element within the irradiated zone. Furthermore, fluid elements will display a distribution of residence times. The dose estimate from PSS cannot provide information on the distribution of doses delivered by a UV system; it can only provide an estimate of the average dose. This is important because two systems with identical average UV doses will not necessarily deliver the same UV dose distribution; nor will they achieve the same degree of microbial inactivation.

The only circumstance under which the average dose may be used to accurately characterize a continuous-flow system is one in which ideal plug-flow conditions are achieved (that is, complete radial mixing and zero longitudinal dispersion). The work of Severin *et al.* (1983 and 1984) suggests that it is difficult to achieve a condition in which complete mixing may be positively assumed. Comparisons of reactor performance (Blatchley and Hunt, 1994) based on the simultaneous application of PSS and bioassay confirm that ideal plug-flow conditions cannot be assumed for purposes of dose estimation in open-channel UV systems. In fact, the observed behavior of these open-channel systems was described more accurately by an ideal completely mixed reactor model than by an ideal plug-flow reactor model.

The bioassay relies on a comparison of batch dose-response data with observed kill in a continuous-flow system. The batch data represent kill in a system from which UV dose can be characterized by a single number. However, the observed kill is taken from a flow-through system, which delivers a distribution of doses. Therefore, the dose estimate from the bioassay provides a measure of the "equivalent" dose required from a system that delivers an ideal (uniform) dose, such as a completely mixed batch reactor or an ideal plug-flow reactor.

Dose estimates from PSS and bioassay should be identical in systems that can be characterized completely by a single dose. In continuous-flow systems, for which this condition will not apply, the PSS dose estimate should represent the upper limit estimate taken by bioassay (Blatchley and Hunt, 1994). Therefore, close agreement between PSS and bioassay dose estimates indicates hydrodynamic efficiency in UV systems.

At present, no technique exists to estimate the dose distribution delivered in a full-scale UV system. Therefore, designers must rely on PSS or bioassays for system analysis. These techniques will provide a measure of system performance in the form of a single parameter: either average dose (PSS) or "equivalent" dose (bioassay). System design should be based on a comparison of "doses" (as defined by PSS *or* bioassay) from pilot investigations with those of successfully implemented UV systems having comparable geometry, hydraulic characteristics, and wastewater quality. It is important that designers understand the meaning of these test results, their fundamental differences, the factors that influence the results, and proper interpretation.

HYDRAULICS

From the perspective of hydraulic behavior, the critical issues in UV system design are promotion of "plug-flow-like" conditions and minimization of head loss. Mixing and head loss behavior will both be governed by the geometry and hydraulic loading of the system. Therefore, the optimal system design should achieve acceptable levels of longitudinal dispersion and head loss.

LONGITUDINAL DISPERSION. A detailed description of longitudinal dispersion measurement techniques and interpretation of test results is given in Chapter 4. Specific attention here will focus on those aspects of dispersion that are unique to open-channel UV systems.

Measurements of dispersive behavior in UV disinfection processes are complicated by the short detention times (< 10 s) typical of these systems. These short detention times dictate extreme precision and accuracy in the stimulus (for example, tracer) introduction and measurement of the corresponding response. In addition, response measurement hardware must provide true real-time measurements of tracer concentration—a lag in the response

---

time of tracer concentration measurements will yield unacceptable results. Consequently, the conventional practice of pulse-injecting a fluorescent dye, commonly done in chemical disinfection systems, cannot be used for UV systems. The pulse-injection technique does not yield reproducible data, and conventional fluorometers do not respond rapidly enough to provide useful response data.

Most successful attempts at measuring longitudinal dispersion in open-channel UV systems have employed a step change in salt concentration, with on-line measurements of the corresponding salt concentration (for example, conductivity) at downstream locations. The salt, usually sodium chloride (NaCl), is introduced at the upstream end of the irradiated zone at a constant rate until conductivity reaches a steady-state value at conductivity cells located immediately downstream of the irradiated zone. Following the achievement of steady state, the salt injection is terminated, and the conductivity (concentration) die-off can be measured. The decay response of the conductivity signal from this step change in salt concentration can be used to estimate the longitudinal dispersivity (E) or dispersion number (d) using the methods described by Levenspiel (1972).

Almost universally, the results of these tests reveal small extents of longitudinal dispersion. Typically, dispersion numbers of less than 0.02 are measured in these systems (Blatchley *et al.*, 1994, and HydroQual, 1992a).

The interpretation of these data is complicated by several factors unique to UV systems. The salt concentrations required for hydrodynamic testing of UV systems can be high enough to induce density currents. As a result, the salt plume may display a vertical component of velocity. This vertical movement can make location of the plume difficult. In addition, conductivity measurements provide an indication of dispersive behavior at a single point in the system. It is likely that longitudinal dispersion is enhanced near the channel walls as a result of boundary shear layers. Therefore, the quantification of dispersive behavior by conductivity measurements should be viewed as a representative value for the entire system. Transverse mixing within UV systems will also affect process efficiency. To date, no quantitative measurements of transverse mixing with UV arrays have been presented.

To a large degree, mixing behavior within the irradiated zone is beyond the control of the designer because of the dominant role of the lamp arrays themselves. In terms of facility design, the most critical factor in minimizing short-circuiting may be the geometry of inlet and outlet structures. These structures should be designed to promote uniform velocity profiles (that is, plug flow) both upstream and downstream of the irradiated zone (see Figure 7.16). In multichannel systems, these structures must also serve the purpose of facilitating uniform flow distribution between channels.

A number of strategies have been used to achieve these performance goals. Inlet flow conditioning is achieved through the application of hydraulic structures, such as stilling plates (see Figure 7.17) and submerged dams. These

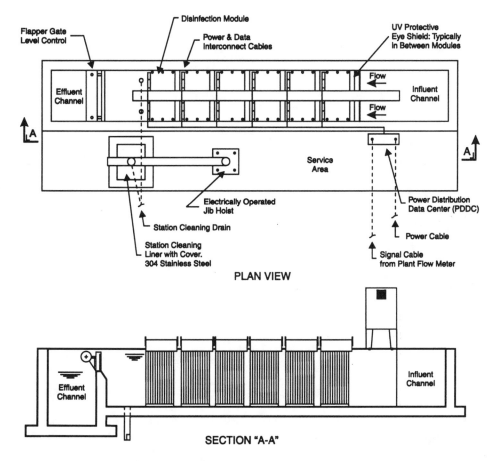

PLAN VIEW

SECTION "A-A"

Figure 7.16   **Profile schematic of lamp modules relative to inlet and outlet structures; lamp modules are placed far enough away from flow structures to ensure uniform flow at the entrance to and exit from the irradiated zone**

structures impose a controlled energy loss on the system influent and are effective in achieving an even distribution of momentum throughout the channel. By positioning the inlet structures far enough upstream of the irradiated zone, flow irregularities (fluid shear) induced by the inlet structure are given ample time for dissipation, thereby allowing a uniform velocity profile to be imposed on the first bank of UV lamps.

A similar logic applies to outlet structures: flow patterns leaving the irradiated zone should be uniform. Outlet structures must also allow liquid level control over the range of expected flow conditions. Several alternatives have been used to achieve these performance objectives, including elongated weirs and flap gates (see Figures 7.17 and 7.18). Flap gates are usually used on larger systems. Elongated weirs have the advantage of no mechanical components and are often used on smaller systems. Elongated weirs are also

---

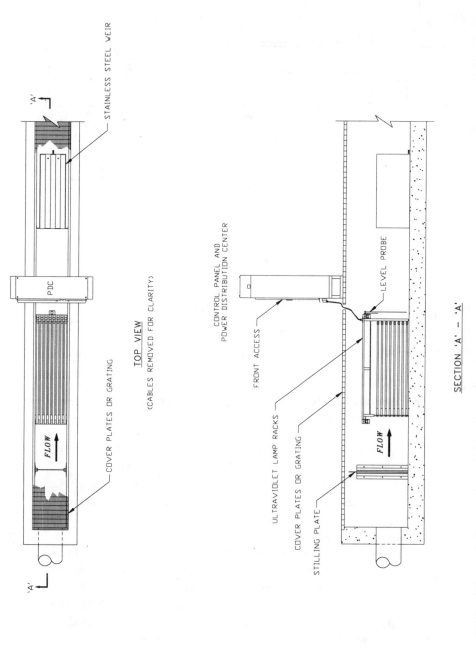

Figure 7.17 **Schematic illustration of ultraviolet disinfection system with stilling plate for flow conditioning and elongated weir for level control**

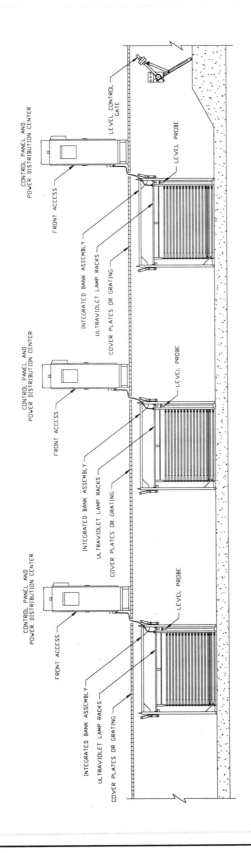

Figure 7.18 Schematic illustration of flap gate and submerged dam used as outlet structure at Bonnybrook wastewater treatment plant, Calgary, Alberta, Canada

potentially advantageous in systems with low overnight flows because they are less likely to allow channel draining than flap gate systems.

As discussed above, the placement of inlet and outlet structures relative to lamp arrays is critical to achieving uniform flow. Measurements of velocity profiles in full-scale systems (Blatchley *et al.*, 1995) suggest that a minimum of approximately 2 m (5 ft) should be allowed between inlet/outlet structures and the closest lamp array. Lamp arrays placed within these inlet/outlet zones may be used ineffectively because of induced abnormalities in the flow structure.

Head Loss. Head loss in open-channel UV systems is manifested as a drop in the water free surface through the system. In terms of power consumption, the energy loss in these systems is inconsequential compared with other losses in a WWTP. However, the drop in the free surface can induce operational problems in the disinfection process. If the liquid level is set such that the downstream free surface is coincident with the top of the irradiated zone, then some liquid on the upstream end of the system will pass through a region of low intensity. Conversely, if the liquid level is set such that the upstream free surface is coincident with the top of the irradiated zone, then a portion of the downstream lamps will not be immersed. With the diurnal fluctuations in flow experienced at most WWTPs, this allows some lamps to experience alternate conditions of immersion and dryness, which can lead to fouling of the quartz jackets surrounding the lamps.

Head loss measurements taken at facilities with existing UV disinfection systems are represented in Figure 7.19. These measurements provide an indication of the head loss that can be expected per UV module at a given approach velocity. Empirical observations in these systems indicate that acceptable performance can be achieved by maintaining total head losses of less than 10 cm (4 in.). In some cases, the effects of head loss can be minimized by construction of a stepped channel (see Figure 7.20). Designers should use caution in adopting this practice for facilities at which wide diurnal flow variations are expected because of the possibility of flooding under low-flow conditions, when head losses are relatively small.

Energy (head) losses in UV systems are a strong function of approach velocity. As in many fluid mechanics applications, a general equation can be written to describe the functional relationship between head loss and velocity:

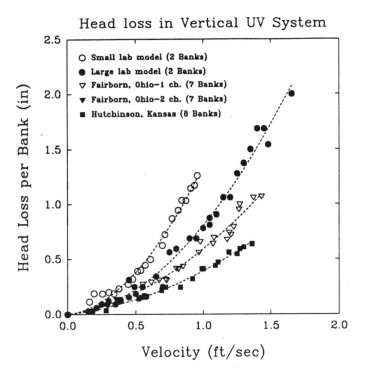

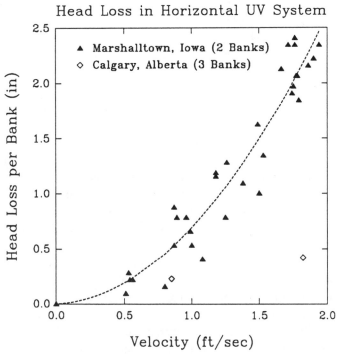

Figure 7.19   Measured values of head loss (per bank) as a function of
approach velocity in open-channel ultraviolet disinfection
systems (Blatchley *et al.*, 1994)

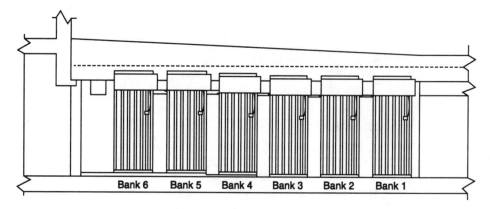

Figure 7.20 Use of a stepped channel to minimize the effects of head loss in an open-channel ultraviolet disinfection system

$$\frac{\Delta H}{L} = aV + b\rho V^2 \qquad\qquad (7.22)$$

Where

ΔH = head loss, cm;

L = channel distance over which ΔH is expressed, cm;

V = approach velocity, cm/s;

ρ = liquid density, g/cm^3; and

a,b = empirical constants.

For a given system geometry, the constants a and b may be determined by experiment. The first term on the right-hand side of Equation 7.22 describes energy losses attributable to viscous forces; the second term describes losses caused by inertial forces. Under laminar flow conditions, only the first term in Equation 7.22 will be important. Under turbulent conditions, only the second term is important. Nonlinear regression of $\Delta H/L$ versus V data for several UV systems indicates that the second term accounts for most of the head loss observed in open-channel UV systems. Therefore, the hydraulic behavior within the irradiated zone can probably be characterized as turbulent, or perhaps transitional, flow. As described in the section on intensity, turbulent conditions are desirable in UV systems. The intensity of turbulence (and associated mixing) will increase with approach velocity.

Longitudinal dispersion in flow-through systems has been characterized using the concept of a dispersion number, d (Levenspiel, 1972). Field measurements indicate that dispersion number d decreases as approach velocity increases, while head loss increases with velocity (Blatchley *et al.*, 1994). Because low values of *d* and ΔH are desirable in UV systems, the selection of an appropriate approach velocity represents an optimization problem. The optimum design condition will correspond to a situation in which head loss is minimized while achieving an adequate intensity of turbulence. "Conventional" systems operated with approach velocities of 5 to 50 cm/s appear to satisfy these criteria.

*F*ACTORS AFFECTING LAMP OUTPUT

For a given set of operating conditions, lamp output will govern microbial inactivation. Although the processes that govern lamp output are largely beyond the control of WWTP operators, a conceptual understanding of these processes and their effects is beneficial.

Ultraviolet output from mercury arc lamps changes as a function of time. In general, lamps begin with a relatively high output power. Lamp output falls sharply in the first 1 000 to 2 000 hours of operation, followed by a more gradual decline up to the point of failure (see Figure 7.21). The recommended operating life of a mercury arc lamp is generally in the range of 7 500 to 8 000 hours; however, lamps have been operated effectively for considerably longer than this. A recent survey of 30 operating facilities revealed that an operating life of greater than 14 000 hours can be expected for low-pressure mercury lamps (U.S. EPA, 1992). The criterion for replacing the lamps was primarily effluent fecal coliform levels. The decision as to when to replace a UV lamp should consider the price of lamp replacement compared to the increased cost associated with operating aged lamps.

System output can be kept relatively uniform if a schedule of staged lamp replacement is implemented. If performed in a logical, orderly manner, staged lamp replacement can provide for relatively consistent UV output within a system.

Lamp wall temperature is also known to affect output, with an optimum level of between 35 and 50°C (U.S. EPA, 1986a). Generally, holding lamp wall temperatures between 45 and 50°C will maintain maximum output from the lamp. This will be a function of the quartz sleeve diameter (that is, the thickness of the air gap between the quartz sleeve wall and the lamp wall), the liquid temperature, and the power driving the lamp. The smaller the quartz diameter, the cooler the lamp will run over the typical liquid temperature operating range (5 to 30°C). Liquid temperatures between 15 and 25°C will

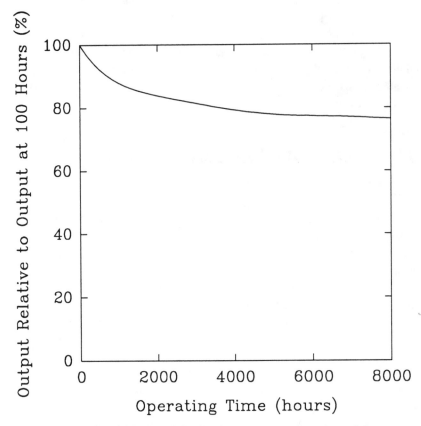

Figure 7.21   Typical ultraviolet lamp output as a function of time

typically result in lamp temperature conditions that are near optimum (greater than 85% maximum output), with outputs falling significantly at liquid temperatures above or below this range.

The electronic ballasts now being installed with all new systems, in lieu of the standard 430-mA electromagnetic ballasts, can provide variable power input to the lamps, which can affect the lamp operating temperature. Manufacturers now offer a range of ballast designs. With constant liquid temperatures, higher currents will drive the lamp temperatures up and vice versa. Thus, the impact of liquid temperature on lamp output can be offset by the ballast input, suggesting that the lamp output can be held near optimum over a wide range of operating conditions. Overall, engineers now have the option of designing for the expected liquid temperature operating range of the WWTP. For example, if the WWTP is located in a warm climate, design considerations can include the choice of quartz sleeve diameter and ballast type that will maintain optimum output over a liquid temperature range of 20 to 30°C.

MATHEMATICAL MODELS

Scheible (1987) presented a model for predicting UV process performance based on the nonideal reactor theory presented in many chemical engineering textbooks (Levenspiel, 1972). A detailed description of the model is presented in *Design Manual: Municipal Wastewater Disinfection* (U.S. EPA, 1986a). The governing equation for the model is as follows:

$$N = N_o \exp\left[\frac{ux}{2E}\left\{1 - \left(1 + \frac{4E(aI_{avg}^b)}{u^2}\right)^{1/2}\right\}\right] + cSS^m \qquad (7.23)$$

Where

N	=	bacterial density after irradiation, cfu/100 mL;
N_o	=	bacterial density before irradiation, cfu/100 mL;
u	=	wastewater velocity, cm/s;
x	=	length of irradiated zone in the direction of flow, cm;
E	=	longitudinal dispersion coefficient, cm²/s;
I_{avg}	=	spatially averaged UV intensity within the irradiated zone as estimated by PSS, mW/cm²;
SS	=	effluent suspended solids concentration, mg/L; and
a,b,c,m	=	empirical coefficients.

The model is rationally based and comprehensive in that it follows conventional process theory and accounts for the factors that are known to affect UV process performance. Specifically, the model has incorporated terms that account for longitudinal dispersion, UV lamp output, and the presence of suspended solids (SS). As such, the model should be able to provide reasonable predictions of changes in process performance resulting from factors such as changes in hydraulic loading, lamp aging, lamp fouling, or variations in effluent SS. Experience with the model suggests the ability to predict effluent coliform density within approximately one order of magnitude (Scheible, 1987). The model has been successfully employed in the design of UV facilities (Gilbert and Scheible, 1993).

A significant drawback to the application of the Scheible model is the need to determine representative values of four empirical coefficients ($a,b,c,$ and m). The parameters a and b are used to relate the average intensity to the rate of microbial inactivation. The parameters c and m are used to relate SS to the concentration of viable, particle-associated microorganisms in the effluent. These coefficients can be determined experimentally but are site-specific.

The strengths of the Scheible model are its rational basis and the inclusion of terms that are measurable, interpretable, and have physical significance in UV systems. All remaining terms in the model can be measured using standard techniques. Appropriate methods for measuring I_{avg} and E in UV systems have been described previously in this chapter.

Although the U.S. EPA model has been used extensively, other models have been developed for predicting disinfection efficacy in flow-through systems. Emerick and Darby (1993) proposed an empirical model of the following form:

$$N = f(\text{dose})^n \qquad (7.24)$$

Where

N	=	effluent coliform concentration, MPN/100 mL;
f	=	water quality factor;
dose	=	average UV dose, as estimated by PSS, mW·s/cm^2; and
n	=	empirical coefficient related to UV dose.

A hypothesis of this model is that inactivation can be predicted from a knowledge of UV dose and a measure of water quality (that is, the water quality factor f). An empirical relationship was postulated to describe the water quality factor:

$$f = A(\text{SS})^a (T)^b (\beta)^c (N_o)^d \qquad (7.25)$$

Where

SS	=	suspended solids concentration, mg/L;
T	=	unfiltered transmittance at 254 nm, %;
β	=	particle size distribution coefficient;
N_o	=	influent coliform concentration, MPN/100 mL; and
A,a,b,c,d	=	empirical coefficients.

A two-parameter power law function is commonly used to describe PSDs in natural waters and treatment systems:

$$\frac{\Delta N}{\Delta d_p} = \alpha d_p^{-\beta} \qquad (7.26)$$

Where

N = particle number concentration, number/cm^3;

d_p = particle size, mm; and

α, β = empirical coefficients.

A comprehensive description of this power law function and its applications is presented by Kavanaugh *et al.* (1980).

The coefficient β provides a measure of the distribution of particles among small and large sizes. Large values of β indicate a PSD that is dominated by small particles, while small values of β are indicative of PSDs that are dominated more by large particles. As described in the background section, particle size plays a key role in shading and shielding microorganisms from UV radiation. Therefore, the inclusion of β as a parameter in the water quality factor appears to be warranted.

Using data from pilot testing at two WWTPs, multiple linear regression analysis showed the parameters N_o and β to be statistically insignificant. The authors hypothesized that the reason for the statistical insignificance of N_o was that the limitation to effluent quality (as measured by N) was not the total number of microorganisms imposed on the system but, rather, the number of particle-associated microorganisms. While dispersed organisms are easily inactivated, those associated with colloidal material are difficult to inactivate by UV irradiation. Furthermore, the inclusion of the SS term in Equation 7.25 was assumed to be capable of accounting for particle-associated organisms.

The data used to perform the regression analysis encompassed wide variations in the SS, T, and N_o values. However, only a narrow range of PSD coefficients was used. Therefore, the effects of β on effluent coliform concentration could not be evaluated with the database used to develop the model.

Given this information, the functional relationship used to describe the water quality factor reduced to the following:

$$f = A(\text{SS})^a (T)^b \tag{7.27}$$

A graphical representation of Equation 7.27 is presented in Figure 7.22. Effluent coliform viability (N) could then be predicted with knowledge of the water quality factor and UV dose (see Figure 7.23). The model was shown to have a somewhat improved ability to predict coliform inactivation compared to the U.S. EPA model. It is important to recognize that this comparison was based on the application of the Emerick and Darby (1993) model to pilot data from only two facilities. The model will be applied at additional facilities in the future, which should allow a more thorough assessment of its capabilities. This extension to other facilities may also allow the inclusion of PSD information in the model.

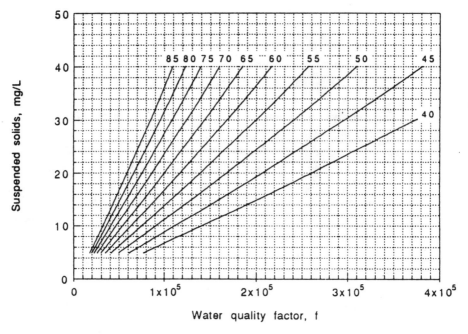

Figure 7.22 Determination of the water quality factor (f) by graphical means; lines of constant transmittance ($\lambda = 253.7$ nm and path length $= 1.0$ cm [10 mm]) are indicated within the figure (Emerick and Darby, 1993)

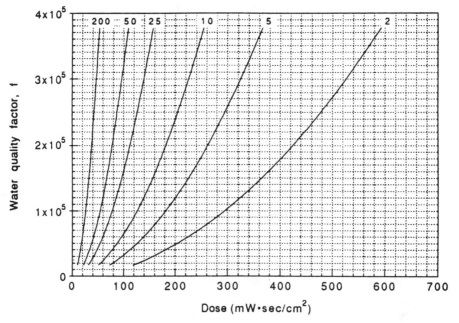

Figure 7.23 Estimation of effluent coliform viability (N) by graphical means; lines within the figure refer to predicted viable effluent coliform concentration (Emerick and Darby, 1993)

Severin *et al.* (1983 and 1984) developed a theoretical model for application to a completely mixed annular UV reactor. The authors performed experiments on a reactor for which a complete assessment of hydrodynamic conditions (that is, complete mix) was available. By applying their knowledge of reactor hydrodynamics and a theoretically derived kinetic model, they were able to achieve accurate predictions of microbial inactivation for several microorganisms (see Figure 7.24).

Unfortunately, this model is of little use for the modular, open-channel systems used in the majority of disinfection operations because of differences in reactor geometry and mixing conditions. However, it is important to recognize the ability of this model to fit actual data from continuous-flow operations. The availability of a complete description of reactor hydrodynamics played a key role in enabling the Severin model to achieve such good predictive behavior. The complex nature of fluid flow through lamp arrays precludes the development of an analogous model for open-channel systems at this time.

It is important to recognize that an extremely large energy input was required to achieve a complete-mix condition in the annular UV reactor (Severin *et al.*, 1984). Open-channel systems do not include provisions for such energy input. Therefore, any assumption of complete mixing (radial or longitudinal) in the modeling of open-channel UV systems would appear to be without justification.

*F*OULING

The ability to deliver radiation from the source to the target is critical to the performance of UV disinfection systems. As discussed previously, dissolved and particulate materials in the liquid phase may impede radiation transmission. Another factor that can limit radiation delivery is the accumulation of insoluble materials on the surfaces of the quartz jackets that house the UV lamps.

Quartz jacket fouling matter can contain organic and/or inorganic constituents. Organic fouling is largely attributable to floatable materials that accumulate on lamp jackets near the free surface in open-channel systems. Control of organic fouling can be achieved by removal of these wastewater constituents in upstream processes.

The inorganic components of a fouling material will accumulate over the entire wetted surface of a quartz jacket. Chemically, these materials are similar to inorganic scale that can form in plumbing or on heated surfaces (such as heating elements). Empirical observations of jacket fouling have suggested that waters containing high hardness and/or high iron concentrations are likely to promote fouling. Unfortunately, relatively little quantitative information exists relative to inorganic scale formation. Therefore, pilot testing has been used extensively to define site-specific fouling potential.

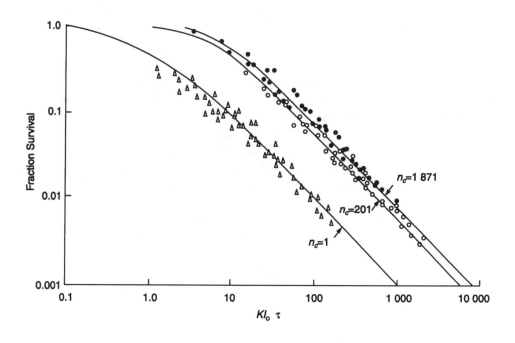

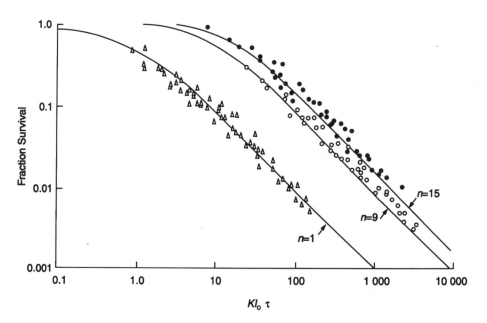

Figure 7.24 Model predictions of effluent viability for *E. coli* (○), *C. parapsilosis* (●), and coliphage f2 (△) from a complete-mix ultraviolet reactor using the multitarget (top) and series-event (bottom) inactivation models (from Severin, B.F., *et al.* [1983] Kinetic Modeling of U.V. Disinfection of Water. *Water Res.*, 17, 1669)

Control of lamp fouling is achieved by a variety of techniques. Chemical removal of scale is achieved by applying a dilute acid (pH of approximately 1 to 2) to the fouled surface. Acid can be applied by either wiping the individual lamps or immersing entire lamp modules. Immersion techniques are probably more efficient for scale removal. For large systems, module immersion hardware is a necessity.

Several different acid solutions have been used for chemical cleaning, including citric acid, phosphoric acid, and commercially available bathroom cleaners. Selection of an appropriate acid will depend on site-specific requirements, but disposal of spent acids should be incorporated in the decision. For large systems, the use of food-grade citric acid or phosphoric acid should be considered so that the neutralized liquid containing the spent acid can be diverted to the headworks of the WWTP.

A number of physical processes can be incorporated to mitigate scaling. Introducing air bubbles at the base of a channel for short periods but on a frequent basis (such as 10 minutes per day) has been shown to effectively mitigate scale formation (Blatchley *et al.*, 1993). This procedure will not eliminate the need for cleaning at facilities where fouling occurs but will be effective in increasing the interval between cleanings.

In a survey of 30 operating WWTPs (U.S. EPA, 1992), a wide variety of cleaning agents was found to be used. Citric acid and commercially available bathroom cleaners were used most commonly. Other agents included commercial detergents and dilute acids. Generally, WWTPs should use a commercially available, inexpensive cleaning agent that is handled and disposed of easily. Materials issues (corrosion, for example) should be considered if cleaning is to be performed *in situ*. A small bench-scale flow-through unit can be used to evaluate a number of agents by trial and error. This would also apply to an evaluation of frequency. Reported cleaning frequencies are highly site-specific and range from weekly to yearly, with a median frequency of approximately once per month (U.S. EPA, 1992).

*G*ENERAL CONSIDERATIONS IN ULTRAVIOLET SYSTEM DESIGN

At present, UV system design relies on a combination of past experience, pilot testing, and numerical modeling. Each factor is related, and the degree to which each is used often depends on the size of the system being considered, the budget, and the schedule. Use of a low-pressure lamp in open-channel configurations is conventional practice today and has received significant experience among many operating WWTPs and the major UV suppliers. In smaller systems, it is probably sufficient to base the design on the expected wastewater characteristics and conventional practices, unless obvious differences from typical municipal wastewater exist. Acceptable numerical models can be used with appropriate default values for design parameters and coefficients. Redundancy should be incorporated into the design, and the sizing should be relatively conservative.

For the design of medium to large facilities, capital and operating costs can be substantial; in such cases, it is important to base the design sizing on relevant and site-specific wastewater characteristics. Pilot testing is recommended, particularly if advanced, nonconventional UV systems are being considered.

DESIGN WASTEWATER CHARACTERISTICS. Data should be collected to obtain a thorough characterization of the effluent quantity and quality. For existing facilities, direct sampling and testing should be conducted and should address seasonal and diurnal variations. If the facility is new, an effort should be made to develop the design effluent characteristics from similar WWTPs and collection systems.

Critical data to be evaluated in design include flow, UV transmittance, SS (preferably including PSD information), and viable indicator organism (for example, coliform bacteria) concentrations. Typically, the UV transmittance of secondary effluents will be greater than 60% on a filtered basis, although lower values, on the order of 50%, have been observed. When collecting UV transmittance data, it is important to measure the parameter on both filtered and unfiltered samples. The filtered measurement presents a more representative estimate of the transmissibility through the effluent water and is critical to the design sizing of the system. The lower the transmittance, the greater the size requirements will be. In some cases, particularly at low transmittance levels, it may be necessary to reduce the spacing of the lamps or consider using advanced higher intensity systems to overcome the lower transmissibility of the water. This is generally the case at transmittance levels of less than 50%.

It should be noted that using the filtered UV transmittance for design will generally yield a conservative (oversized) design. A correlation presented in U.S. EPA's *Design Manual: Municipal Wastewater Disinfection* (U.S. EPA, 1986a) provides a more representative corrected UV absorbance value; this would yield a smaller design sizing, a consideration that would significantly affect large-system designs.

Suspended solids will affect the transmittance of UV, occlude bacteria, and generally interfere with the UV disinfection process to a greater degree than encountered with chemical disinfection systems. This, in effect, establishes a limit of disinfection efficiency that can be accomplished by UV; this limit is a function of the particulate matter in the effluent. Certainly, one can expect this effect to vary and be dependent on the type of particulate matter and size distribution of the particles. In waste streams containing a high degree of inorganic matter or naturally occurring soil solids (such as in combined sewer overflow and stormwater), the concentrations of viable organisms associated with particulate matter may be relatively small. In cases of typical municipal wastewater and biologically treated wastewater, however, these concentrations can be significant and account for essentially all residual coliforms in the final effluent after clarification. For this reason, a high degree of filtration is required, including, in some cases, chemical coagulation of colloidal solids, to achieve high disinfection efficiencies. This is the case with California Title 22 reuse requirements, in which total coliform levels less than 2.2 cfu/100 mL are targeted.

Scheible (1987) suggested a correlation with SS to predict the level of particulate coliform after UV disinfection of treated municipal effluents:

$$N_p = c \, SS^m \tag{7.28}$$

Where

N_p	=	particulate coliform density, and
c, m	=	coefficients representing intercept and slope, respectively, of a log–log regression analysis of SS, with the effluent coliforms measured after imposition of high UV doses.

This is part of the model described by U.S. EPA (1986a). HydroQual (1994) further compiled fecal coliform and SS data from various studies and conducted a regression analysis as described above. A linear regression (transformed) yielded the following expression:

$$N_p = 0.69 \, SS^{1.6} \tag{7.29}$$

With a correlation coefficient (r^2) of 0.67. These data are presented in Figure 7.25. The variation is wide, but the correlation serves as a useful screening tool to estimate the fecal coliform associated with particulates in a secondary effluent.

The initial bacterial concentration is a critical design factor. In most cases, systems are designed on the basis of indicator organisms such as total and fecal coliform, *Escherichia coli*, and enterococcus. System sizing, or applied "dose" (usually as defined by PSS), is a direct function of the initial concentration. These, in turn are dependent on the degree of treatment before disinfection. Fecal coliform concentrations in secondary effluents are typically on the order of 10^4 to 10^6, while tertiary treatment will result in a lower order of magnitude and primary treated effluent a higher order of magnitude. Because expected initial organism concentrations cannot be predicted solely from the type of treatment process preceding disinfection, the above numbers should be used only as guidelines. Direct testing should be conducted to more accurately characterize the initial concentrations for a specific site.

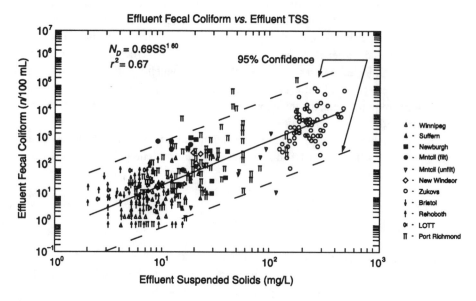

Figure 7.25 **Regression of observed effluent fecal coliform concentration versus suspended solids concentration compiled from several studies of ultraviolet disinfection systems (HydroQual, 1994)**

Development of the design wastewater characteristics must take into account the variability of the relevant water quality parameters and the targeted effluent goals. If the facility permit requirements are stated in terms of maximum daily, 7-day, and/or 30-day averages, for example, the system design parameters must be developed on these bases also. Designing to meet a 30-day maximum average effluent coliform on the basis of maximum daily influent characteristics will only result in a significantly oversized system. Additionally, it is important to note that maximum values (daily, 7-day, 30-day, and so on) for the relevant water quality parameters and flow are generally coincident. This was demonstrated in studies conducted using large databases from two facilities (HydroQual, 1992b). Thus, one can expect that high-flow periods will be accompanied by high solids and coliform concentrations and low UV transmittances. This is a significant factor when considering the design sizing of a UV system.

As a further characterization for the design of a UV system, a collimated beam study of microbial dose-response behavior is valuable. This test, which is relatively inexpensive to perform, involves the exposure of nondisinfected effluent samples to a range of known UV doses and quantification of the inactivation response. If possible, the test should be repeated several times during periods of good and poor effluent quality. The results of these dose-response experiments can be compared with those of previous efforts to further characterize the potential for successful implementation of UV disinfection and possible process limitations. These data can also be related directly to applied dose correlations that exist for specific UV equipment configurations and used to estimate approximate sizing requirements.

PILOT TESTING. If actual pilot testing is to be performed, it is important to ensure that the proper information and performance data are generated from the study. The pilot system itself should be selected to mimic as closely as possible the expected full-scale design, both hydraulically and in lamp loading. Given the modular nature of UV systems, this is a relatively straightforward procedure. For conventional low-pressure lamp systems, one should consider at least two banks in series and 10 to 20 lamps per bank. For high-intensity lamp systems, one should employ multiple lamps or at least a representative module.

The purpose of the piloting should be clearly defined. If process design information and performance data are to be developed, the unit should be operated over a wide range of conditions extending below and above the expected operational conditions of the full-scale system. The unit should be cleaned (quartz surfaces) throughout the testing; the performance data and process evaluation should not be biased by any fouling of the quartz. The pilot test program should be designed to reflect the analysis method planned for the data. For this purpose, modeling techniques are useful because data may be generated to calibrate and verify the model. The data generated should thus be clearly defined to respond to the requirements of the design model. For example, if the Scheible (1987) model is to be employed, then hydraulics, influent and effluent bacteria, and influent transmittance and SS data are all important. The model requires data collection over a range of applied dose levels from low to high. Once the model is calibrated, it serves as a useful tool for estimating system sizing under expected design conditions and can present a perspective of the sensitivity of the design to specific parameters.

For process purposes, the pilot studies can be conducted over a relatively short period of time. Data can be collected through a fairly intensive sampling and analysis effort that includes variations in operating conditions within a short period of time. For example, the unit may be sampled at four dose levels in one day and continued for several days per week until sufficient data are generated to develop the design parameters. This may take 3 to 6 weeks, depending on the level of effort; it is not necessary to operate the unit on a continuous basis. Such an effort might be repeated again if significant seasonal variations are exhibited in the wastewater.

If the purpose of the pilot testing is to gain operating information, particularly with respect to fouling potential, then it is necessary to continue the study for a period of several months. In such a case, the unit should be operated continuously and monitored for quartz sleeve fouling. This may be done by direct observation of the quartz (via a radiometer) or by monitoring coliform reduction under constant loading conditions. Piloting for energy use, labor requirements, or system maintenance is not necessary. Sufficient information exists to estimate these factors, and direct calculations of power based on the ballast/lamp configuration are more than adequate.

An effective pilot design may also involve direct testing to estimate the actual dose delivered as a function of hydraulic loading (L/min/W, for example). These tests can be done via the bioassay approach by correlating the data with direct dose-response information developed from collimated beam studies conducted for the specific application. Procedures exist to conduct this work either in clean water (generally using *Bacillus subtilis* spores) or in the actual wastewater matrix (using an MS2 coliphage, for example). This allows a direct quantification of dose and extrapolation to system sizing requirements, expressed as the number of lamps required to meet specific dose levels.

SYSTEM SIZING AND CONFIGURATION CONSIDERATIONS.

Final design of a full-scale UV system will include establishing the number of lamps required to meet disinfection requirements under design conditions. Equally critical is the manner in which they are configured in the full-scale design. Typically, lamp banks are arranged in series, usually two or three for horizontal lamp systems and three to six for conventional vertical lamp systems. Furthermore, it is preferable to design the system with relatively long, narrow channels to encourage plug flow and avoid any degree of short-circuiting. As a screening guideline, an average design sizing of 37 conventional 1.5-m lamps per million gallons per day (mgd × $[3.785 \times 10^3] = m^3/d$) of peak design flow may be used (U.S. EPA, 1992). Certainly, this average is based on a wide range, and any specific WWTP size will be dependent on site conditions and effluent quality.

Hydraulic design is one of the more critical factors to consider when laying out the full-scale system. An ineffective hydraulic design can cause failure of the system to meet disinfection requirements. Closed-shell systems using conventional lamps, which were widely used in earlier UV installations, often experienced performance problems because of poor hydraulic behavior. The current practice of designing open channels that are long and narrow mitigates these concerns.

In designing the channel to house the UV modules, it is important to include proper inlet and outlet structures and consider approach and exit conditions. Upstream, a perforated stilling plate can be installed if sufficient head is available. This distributes the flow and equalizes the velocities across the cross section of the channel. The stilling plate should be placed at least 1.5 m (5 ft) in front of the first lamp bank. Otherwise, the channel should have an undisturbed straight-line approach of two to three lamp lengths. There should be sufficient distance allowed between lamp banks (0.5 to 1 m [2 to 4 ft]) and two to three lamp lengths between the last bank and the downstream level-control device.

Proper design practice, particularly for large systems, entails the consideration of multichannel configurations. Under these circumstances, the inlet structure must satisfy the dual requirements of inducing uniform flow and allowing even distribution of flow among operational channels. Channel inlet structures should also allow for hydraulic isolation of individual channels during low flow and routine maintenance. Operationally, the multichannel design should be controlled to maintain a minimum velocity through any one channel.

In conventional low-pressure lamp systems, wastewater within the channel must be maintained at a constant level, with little fluctuation. Most designs use a mechanical, counterbalance gate downstream of the lamp batteries. These are successful when operated within a specific flow range. Problems have been experienced, however, during times of low flow or no flow. These systems are most appropriate at WWTPs in which these conditions can be avoided; larger, multichannel systems are applicable when the proper flow range can be maintained by opening and closing channels as needed. In smaller WWTPs, fixed or adjustable weirs may be more appropriate. Sufficient weir length must be provided to avoid excessive water level fluctuation.

System control should be a function of the system type and the size of the WWTP. Controls should be simple: the objective is to ensure that system loading can be maintained and disinfection accomplished while conserving the operating life of the lamps. This becomes increasingly important in larger systems. In smaller systems, it may be best to have the full unit in operation at all times, excluding the redundant units incorporated into the design. Manual control and flexibility should be available as the system increases in size, enabling the operator to bring portions of the system (such as channels and banks) into and out of operation as needed to adjust for changes in flow or water quality. Automation of this activity is increasingly beneficial as the system becomes larger and incorporates multiple channels.

Safety, centering on electrical hazards and protection from exposure to UV radiation, is important in the design and operation of the UV process. Exposure risk is generally minimal as long as the operating lamps are submerged and the lamp batteries are shielded. It is generally not necessary to operate lamps in air except under extraordinary circumstances. In no case should the advanced high-intensity lamps be operated in air and unshielded. All systems must be equipped with safety interlocks that shut down the modules if they are moved out of their operating positions or the wastewater level falls, leaving any or all lamps exposed to air. Electrical hazards are minimized by the inclusion of ground-fault-interruption circuitry with each module. This should be a specified feature of all systems.

An overriding concern in the proper maintenance of the UV reactor is to keep all surfaces through which the radiation must pass clean and fully transparent to the UV radiation. Prevention of surface fouling is critical: insufficient cleaning can often be the primary reason for improper performance of a particular system. Proper design should include easy access to the lamp modules for cleaning and other maintenance tasks. The installation should consist of a large enough area for working conditions and for handling the modules when taken out of the channels.

Various methods of cleaning are provided with current UV systems. For small, conventional low-pressure units, manual cleaning is sufficient. Dip tanks and racks for the individual module should be provided for manual cleaning. In larger systems, the modules are removed in banks and cleaned in a dip tank; in this case, a traveling hoist is needed for removing and handling the modules. Mechanical wipers, with and without chemical cleaning capability, are provided with some systems.

The reactors, channels, and related tankage should be equipped with drains to allow for complete and rapid dewatering; drainage should be directed back to the headworks of the WWTP. A clean-water system should be permanently available for rinsing and cleaning needs. Consideration should be given to providing a bypass around the UV system, particularly in WWTPs that have seasonal disinfection requirements.

Screening should be considered upstream of the UV units to remove any debris from the wastewater. Algae, in particular, have caused problems when sloughing from the upstream clarifiers and channels. Leaves and plastic debris have also been observed. These materials tend to catch on the lamps and cause difficulties. Cleaning can present a maintenance problem. The screens can range from simple mesh inserts that are manually removed and maintained to self-cleaning, mechanical moving screens.

RETROFIT CONSIDERATIONS. Many WWTPs are abandoning chlorination and switching to UV disinfection. Existing chlorine contact chambers offer an opportunity to cost-effectively install the equipment. The channels are simply modified with a false floor and interchannel walls to accept the equipment. Often, only a portion of the contact chamber is needed for this purpose, while the remaining portion can be used for future expansion.

The most significant hydraulic constraint often encountered in retrofit applications is the available hydraulic head; this factor should be carefully considered in the design of the system. Additionally, chlorine contact tanks are relatively wide. They should be split into multiple channels to provide a high length-to-width ratio conducive to plug flow.

CURRENT ULTRAVIOLET EQUIPMENT

Original systems offered by vendors in the early 1980s consisted of enclosed chambers employing either a submerged-lamp system or a noncontact lamp system. The technology evolved to a modular, submerged-lamp system installed in an open channel, which significantly improved system maintenance and afforded better hydraulics. The modular, open-channel UV system using a conventional low-pressure mercury arc vapor lamp is currently the industry standard. An estimated 80% of all UV systems in operation today are open-channel, low-pressure lamp systems, and they constitute nearly all of the recent and new installations. A recent major improvement to these systems was the development of the electronic ballast, which has been available for approximately 5 years. The current emphasis in the market is research and development of alternate high-intensity UV sources, which fall into two basic categories: high-intensity low-pressure lamp systems and medium-pressure lamp systems. Changes in lamp physics allow each of the new systems to provide similar germicidal performance, with substantial reductions in the number of lamps used (one-eighth to one-twentieth the number of lamps) compared to conventional low-pressure lamp systems.

LOW-PRESSURE MERCURY LAMP SYSTEMS. The low-pressure mercury arc lamp principle is employed in germicidal and standard fluorescent lighting lamps. Both produce UV radiation by means of an electric discharge through a mixture of mercury vapor and argon at a controlled subatmospheric pressure (0.007 mm Hg [torr]). For ultraviolet lamps, this occurs in a transparent tube, while fluorescent lamps use a phosphor-coated tube that converts UV light to visible light.

The low-pressure mercury vapor lamp is the most common lamp used for wastewater disinfection. It has the longest performance history of the three major lamp types. This lamp has been the industry standard since the introduction of UV disinfection systems and accounts for more than 99.9% of the UV installations in the U.S. and Canada.

Two standard lamp lengths are typically used in conventional disinfection systems: the 36-in. (30-in. arc length) (0.9-m [0.7-m arc length]) and the 64-in. (58-in. arc length) (1.6-m [1.5-m arc length]). Both are commonly employed in vertical lamp systems, while the 64-in. (1.6-m) lamp is typically used in horizontal lamp systems.

The introduction of electronic ballasts to drive UV lamps was a significant improvement to conventional low-pressure lamp systems. The older, conventional electromagnetic ballasts, although reliable, were inefficient and susceptible to overheating, and did not allow modulation of the power supplied to the lamps. The electronic ballasts, which have become standard and are used in most new systems today, are solid state. The electronic ballasts are significantly lighter, more compact, modular (plug-in design), energy efficient, and allow modulation of the power supply to the lamps. The electronic ballast's ability to dim lamps allows better and more cost-effective flow pacing of the UV system. Specification and selection of electronic ballasts are not standard, with variations offered by different manufacturers. The alternate lamp systems have not, as a rule, had these ballasts incorporated into their system designs, although such systems are being developed, and one medium-pressure lamp system has recently been introduced with an electronic ballast.

While the low-pressure lamp is efficient at producing effective germicidal radiation, its output intensity is relatively low. The UV output is 0.18 W of UV per centimetre of arc length (0.46 W/in.). This yields standard outputs (at 254 nm) from new 36-in. (30-in. arc length) (0.9-m [0.7-m arc length]) and 64-in. (58-in. arc length) (1.6-m [1.5-m arc length]) lamps of 13.8 and 26.7 W, respectively. Systems require relatively large numbers of these lamps in fairly densely packed lamp banks (2- to 5-in. [50- to 125-mm] spacings).

Low-pressure lamp systems have become increasingly reliable from both an operations and a performance standpoint, to the point where reliability is generally no longer a factor in comparison with other disinfection technologies. Lamps are widely available at a relatively low cost. Initially, and with the first-generation closed-shell reactors, lamp life was estimated to be approximately 7 500 hours; today, with actual operating experience and the advances in open-channel, full-submergence systems, effective lamp lives have been shown to be greater than 13 000 hours. This results in an 85% longer relamping interval than originally needed, yielding a significant reduction in operating costs.

Low-pressure lamp systems are available in several open-channel modular configurations. However, closed-shell and noncontact low-pressure lamp systems are no longer manufactured for wastewater applications. Open-channel systems fall into two major categories: horizontal and vertical.

HORIZONTAL ULTRAVIOLET SYSTEMS. Open-channel, modular, horizontal UV lamp configurations are the most prevalent systems in the municipal wastewater industry. In 1990, more than half of all systems in operation were horizontal. When open-channel systems alone were considered, they outnumbered vertical systems eight to one. Although alternate lamp systems are receiving increasing consideration (especially at large installations), the low-pressure lamp configurations are usually specified, and the ratio of horizontal-to-vertical lamp placements for new systems may currently be more than 10:1.

Horizontal lamp systems consist of lamp bundles that are suspended from modular racks in planes parallel to the channel floor. Most suppliers in this category provide systems with lamps that are parallel to the direction of process flow. Lamp bundles, referred to as *banks*, consist of a number of modules that span the channel width. Because of its modular nature, a bank of lamps may contain any number of modules. The module consists of a metal support frame through which the lamp wiring runs to any number of evenly spaced quartz-jacketed lamps. In large systems, the modules typically hold either eight or 16 lamps, and smaller systems hold as few as four to six lamps per module. Large systems are offered with UV banks mounted in "cages" so that a whole bank can be removed for cleaning. Conversely, in most smaller systems, the individual modules are removed for cleaning or maintenance. Horizontal systems are generally of multibank and multichannel design. This allows the economic use of semiautomatic to fully automatic flow pacing and provides system flexibility to allow cleaning and maintenance tasks without a loss in system performance.

The UV lamp is housed in a quartz tube in either a double open-ended tube or a single open-ended test-tube-like shell. The lamp/quartz assembly is secured to the module rack by an o-ring and socket connector. Today's systems are designed with individually isolated lamps; this maintains system integrity in the event of individual lamp failure or breakage. Currently, the industry standard lamp spacing is 3 in. (75 mm) arranged in a uniform lamp array. Early systems were supplied with lamp spacings varying from approximately 1.5 to 4 in. (35 to 100 mm).

Liquid level control is an important concept in horizontal systems. Level-control devices currently in use are designed to maintain a target level within approximately 0.25 in. (6 mm). The target level is generally the height to the top lamp plus half the height of the lamp spacing. This promotes the distribution of a relatively uniform dose to all fluid elements being treated. The level-control device also prevents the liquid level from dropping below the top set of lamps, which could result in both safety and operating problems. The most common liquid-level-control device is the counterbalanced flap gate; fixed and motorized weirs have also been used.

Lamp cleaning of the horizontal system is accomplished by either bank or module removal to a mobile or dedicated cleaning station. The level of cleaning complexity can range from a drained area equipped with a holding rack, hose, and cleaning solution to automatic air sparging or an ultrasonic dip tank for large banks accessed with overhead hoists.

System control varies from minimal to fully automatic. Fully automatic systems enable system control from a remote location such as a central operations center. System controls usually provide, at minimum, system power, system hours, and lamp status indicators. Fully automatic designs can integrate flow and wastewater conditions and pace the UV system by either dimming lamps, shutting down banks, or taking channels out of service.

VERTICAL ULTRAVIOLET SYSTEMS. Open-channel, modular, vertical UV systems have been operating in the municipal wastewater field since 1987. Vertical systems were brought to the market as an alternative to modular, horizontal, open-channel systems, which saw their first full-scale operation at a WWTP in Canada in 1982.

Vertical lamp systems consist of lamp bundles that are secured in an open rectangular frame. The frame rests on the channel bottom in an upright position (lying on one of its short faces), such that the lamps are perpendicular to the channel floor. A vertical lamp system module typically consists of

40 lamps mounted in a frame unit in an eight-by-five lamp array. Traditionally, these modules have employed a staggered lamp array, in which alternating rows of lamps are parallel to one another but are essentially "out of phase" by one-half of the lamp spacing distance. In theory, this design should result in increased radial turbulence with minimal added axial turbulence. More recently, vertical system manufacturers have been using uniform lamp arrays.

Lamp modules may be placed side by side and/or front to back to form banks. The modules require an overhead crane for removal from the channel. An important feature is that the unit can be relamped with the module in place, unlike the horizontal lamp modules. However, the entire module would necessarily need to be deenergized to permit safe servicing. Vertical systems generally use the shorter 0.9-m (36-in.) lamps, although the 1.5-m (60-in.) lamps have been used for larger systems. The lamp length sets the required liquid depth, which is substantially deeper than used with the horizontal systems.

Lamp cleaning is generally accomplished in a similar manner to that of the horizontal systems. Early systems offered two in-place cleaning systems, one involving the introduction of a cleaning solution into the channel followed by subsequent agitation, the other employing a mechanical wiping system. The wiping system could be used under process operation, while the chemical cleaning system would require taking the unit off line to accomplish lamp cleaning. Current options include an air-scouring system that is engaged in place and under process conditions. It is used to increase the interval between chemical lamp-cleaning cycles, which can either be done *in situ* (isolating the channel) or by transferring the module to a dip tank.

Liquid-level control and system monitoring and control are similar to those found in the horizontal lamp systems. Fixed and motorized weirs are essentially the same, although the tendency toward deep, generally narrower channels would require longer fixed weirs and more active motorized weirs. The difference in counterbalanced flap gate systems is that a base wall is generally provided. Early vertical systems afforded better flow-pacing potential because lamp rows could be turned off without reducing the areal dose of UV radiation. To maximize this advantage, vertical system manufacturers offer rapid-start lamps that allow more frequent on–off cycles than the instant-start lamps used in horizontal systems. Horizontal system flow pacing required shutting down whole lamp banks to effect energy savings. Current systems offer electronic ballasts that allow lamp dimming. Lamp dimming improves flow-pacing ability in both vertical and horizontal lamp systems.

MEDIUM-PRESSURE MERCURY LAMP SYSTEMS. Medium-pressure lamps employ the same basic principle as low-pressure lamps. The major difference is that the mercury vapor emission is carried out at significantly higher lamp pressures and temperatures. The medium-pressure lamp operates in the 10^2 to 10^4 torr (mm Hg) range, which is at or near atmospheric pressure. Lamp operating temperatures range from 600 to 800°C, which is 10 to 20 times higher than the standard operating temperature range of 40 to 60°C for low-pressure lamps. Unlike with the low-pressure lamp, the wastewater temperature has no impact on the medium-pressure lamp operating temperature.

Physical differences include a thin molybdenum foil that connects the electrodes and external connections, and external coating of the lamp ends with a reflective, heat-resistant material. The external coating is used to maintain lamp temperature, thereby preventing mercury condensation. This is critical because, unlike the use of low-pressure lamps in which only a portion of the mercury is vaporized, all of the mercury in a medium-pressure lamp is vaporized. The pressure remains constant and is fixed by the amount of mercury in the lamp.

The UV output of a medium-pressure lamp is 50 to 80 times higher than the output of a low-pressure lamp. Ultraviolet output is typically on the order of 9.1 to 14.2 W/cm arc length (23 to 36 W/in.). However, the radiation produced is polychromatic and ranges from the lower end of the germicidal range (200 nm) to red visible light (approximately 700 nm). While the 30 to 40% conversion of input energy to radiation is similar to that of low-pressure lamps, only approximately 25% of the energy is in the germicidal range. The net effect is that the conversion of input energy to germicidal energy is 5 to 7% for medium-pressure lamps, compared to 30 to 35% for low-pressure lamps.

The typical arc length of a medium-pressure lamp is roughly one-fifth that of the standard 64-in. (58-in. arc length) (1.6-m [1.5-m arc length]) low-pressure lamp. When accounting for the shorter lamp length, higher intensity, and lower conversion to germicidal energy, the theoretical UV output is 8 to 16 times greater than that of a low-pressure lamp.

Medium-pressure lamps have a rated life of 4 000 hours, although experience has shown an expected life exceeding 8 000 hours. The actual lamp life is dependent on lamp operating power. A higher operating power results in higher lamp temperatures and lower lamp life. Because of their currently limited market, the lamps are significantly more expensive, and their availability is limited (from manufacturers only).

The major advantage of the medium-pressure system is the lower capital cost of installation. The expense of facility requirements for a medium-pressure system is 20 to 10% that of a low-pressure system. The cost savings are realized through reduced construction and installation costs. Equipment costs vary from marginally lower to marginally higher, with medium-pressure lamp costs ranging from $300 to $500 per lamp. As the number of applications increases, lamp price discounts can be expected. A second advantage is the decreased requirement for lamp cleaning resulting from the significantly reduced number of lamps. Additionally, manufacturers of these systems provide automatic lamp-cleaning systems that further reduce cleaning efforts.

The major disadvantage is the high operation and maintenance costs (exclusive of lamp cleaning). These systems cost more to operate from an energy standpoint because of their inefficient energy conversion. Maintenance costs relating to lamp replacement are high. Actual relamping costs are marginal, but the medium-pressure lamp replacement cycle is 10 to 40% shorter than that of conventional low-pressure lamp systems. However, when relamping labor is considered, the cost difference may well be minimal.

Experience with medium-pressure lamp systems is limited. Although one WWTP has been operating since 1987, only eight installations are operating today, with at least four more under construction, representing approximately 0.1% of the WWTPs in operation. However, these systems are being considered for installation in ever-increasing numbers.

Two systems are currently offered. Both provide automatic in-place cleaning systems. One supplies dual cleaning consisting of a mechanical wiping system and a chemical in-channel cleaning system. The mechanical wiping system provides routine physical cleaning, while the chemical cleaning system accomplishes more effective cleaning that is required less frequently. The chamber is taken out of service during chemical cleaning operations. The second system incorporates mechanical and chemical cleaning in one unit. It operates while the system is in operation without affecting disinfection performance. This is accomplished by a 50-mm (2-in.) wiper mechanism that circulates cleaning solution under pressure within the wiper as it moves along the lamp length.

Lamp replacement in the second system requires removal of the lamp module from the reactor. This would require taking the unit out of service. Lamp replacement in the first system can be accomplished with the system on line. Lamps are accessed through watertight ports in the chamber wall, with all electrical connections made outside the chamber. Monitoring and control of medium-pressure lamp systems are similar to those employed in low-pressure systems. The lamps have more than one power setting, which allows added flow-pacing capability and increased lamp life.

LOW-PRESSURE, HIGH-INTENSITY SYSTEMS. The aim of the low-pressure, high-intensity lamp is to incorporate the beneficial features of the conventional low-pressure and medium-pressure lamp systems: specifically, the nearly monochromatic germicidal light produced by conventional low-pressure lamps and the high-intensity levels characteristic of medium-pressure lamps. The low-pressure, high-intensity lamp uses a high-current discharge technique that allows operating pressures in the 10^{-2} to 10^{-3} torr range. The actual operating pressure is as much as 40% higher than that of its conventional counterpart. Operating temperatures for high-intensity lamps are in the 180 to 200°C range, five times higher than those of conventional lamps. The high-intensity lamp is driven by currents as high as 5 amps, 10 to 15 times higher than those of conventional low-pressure lamps.

REFERENCES

Blatchley, E.R., III, and Hunt, B.A. (1994) Bioassay for Full Scale UV Disinfection Systems. *Water Sci. Technol.*, **30**, 4, 115.

Blatchley, E.R., III, *et al.* (1993) Large-Scale Pilot Investigation of Ultraviolet Disinfection. *Proc. Plann., Des. & Oper. Effluent Disinfection Syst. Spec. Conf.*, Water Environ. Fed., Whippany, N.J., 417.

Blatchley, E.R., III, *et al.* (1994) Macro-Scale Hydraulic Behavior in Open Channel UV Systems. Paper presented at 67th Annu. Conf. Water Environ. Fed., Chicago, Ill.

Blatchley, E.R., III, *et al.* (1995) UV Pilot Testing: Intensity Distributions and Hydrodynamics. *J. Environ. Eng.*, **121**, 258.

Cairns, W.L. (1991) *Ultraviolet Disinfection: An Alternative to Chlorine Disinfection.* Trojan Technologies, Inc., London, Ont., Can.

Calmer, J.C., *et al.* (1994) Dynamics of Coliform Regrowth in a Dechlorinated Secondary Effluent. Paper presented at 67th Annu. Conf. Water Environ. Fed., Chicago, Ill.

Collins, H.F., and Selleck, R.E. (1972) *Process Kinetics of Wastewater Chlorination.* SERL Rep. 72-5, Univ. of Calif., Berkeley.

Craun, G.F. (1988) Surface Water Supplies and Health. *J. Am. Water Works Assoc.*, **80**, 2, 40.

Darby, J.L., *et al.* (1993) Ultraviolet Disinfection for Wastewater Reclamation and Reuse Subject to Restrictive Standards. *Water Environ. Res.*, **65**, 169.

Davidson, J.N. (1969) *The Biochemistry of the Nucleic Acids.* 6th Ed., Methuen & Co., London, U.K.

Emerick, R.W., and Darby, J.L. (1993) Ultraviolet Light Disinfection of Secondary Effluents: Predicting Performance Based on Water Quality Parameters. *Proc. Plann., Des. & Oper. Effluent Disinfection Syst. Spec. Conf.*, Water Environ. Fed., Whippany, N.J., 187.

Gilbert, S. and Scheible, O.K. (1993) Assessment and Design of Ultraviolet Disinfection at the LOTT Wastewater Treatment Plant, Olympia, WA. *Proc. Plann., Des. & Oper. Effluent Disinfection Syst. Spec. Conf.,* Water Environ. Fed., Whippany, N.J., 137.

Harm, W. (1975) *Molecular Mechanisms for Repair of DNA, Part A.* P.C. Hanawalt and R.B. Setlow (Eds.), Plenum, New York, N.Y.

Harris, G.D., *et al.* (1987) Ultraviolet Inactivation of Selected Bacteria and Viruses with Photoreactivation of the Bacteria. *Water Res. (G.B.)*, **21**, 692.

Helz, G.R., and Nweke, A.C. (1995) Incompleteness of Wastewater Dechlorination. *Environ. Sci. Technol.*, **29**, 1018.

Hunt, B.A. (1992) *Ultraviolet Dosimetry Using Microbial Indicators and Theoretical Modelling.* M.S. thesis, School Civ. Eng., Purdue Univ., West Lafayette, Ind.

HydroQual, Inc. (1992a) *A Review of UV Disinfection Process Design Consideration*. Preliminary draft, U.S. EPA, Washington, D.C.

HydroQual, Inc. (1992b) *Users Manual for UVDIS Version 3.1. UV Disinfection Process Design Manual*. Second draft, U.S. EPA, Washington, D.C.

HydroQual, Inc. (1994) *Disinfection Effectiveness of Combined Sewer Overflows*. Section draft report prepared for Metcalf & Eddy, Inc., U.S. EPA, Washington, D.C.

Jacob, S.M., and Dranoff, J.S. (1970) Light Intensity Profiles in a Perfectly Mixed Photoreactor. *Am. Inst. Chem. Eng. J.*, **16**, 359.

Jagger, J. (1967) *Introduction to Research in Ultra-Violet Photobiology*. Prentice-Hall, Inc., Englewood Cliffs, N.J.

Kavanaugh, M.C., *et al.* (1980) Use of Particle Size Distribution Measurements for Selection and Control of Solid/Liquid Separation Processes. In *Particulates in Water*. Advances in Chemistry, No. 189, Am. Chem. Soc., Washington, D.C.

Lehrer, A.J., and Cabelli, V.J. (1993) Comparison of Ultraviolet and Chlorine Inactivation of F Male-Specific Bacteriophage and Fecal Indicator Bacteria in Sewage Effluents. *Proc. Plann., Des. & Oper. Effluent Disinfection Syst. Spec. Conf.*, Water Environ. Fed., Whippany, N.J., 37.

Levenspiel, O. (1972) *Chemical Reaction Engineering*. 2nd Ed., John Wiley & Sons, Inc., New York, N.Y.

Lindenauer, K.G., and Darby, J.L. (1994) Ultraviolet Disinfection of Wastewater: Effect of Dose on Subsequent Photoreactivation. *Water Res. (G.B.)*, **28**, 805.

March, J. (1985) *Advanced Organic Chemistry*. 3rd Ed., John Wiley & Sons, Inc., New York, N.Y.

Meulemans, C.C.E. (1987) The Basic Principles of UV-Disinfection of Water. *Ozone Sci. Eng.*, **9**, 299.

National Water Research Institute (1993) *UV Disinfection Guidelines for Wastewater Reclamation in California and UV Disinfection Research Needs Identification*. Fountain Valley, Calif.

Oliver, B.G., and Cosgrove, E.G. (1975) The Disinfection of Sewage Treatment Plant Effluents Using Ultraviolet Light. *Can. J. Chem. Eng.*, **53**, 170.

Putnam, L.B., *et al.* (1993) Pilot Testing UV Disinfection on Secondary Effluent at CCCSD. *Proc. Plann., Des. & Oper. Effluent Disinfection Syst. Spec. Conf.*, Water Environ. Fed., Whippany, N.J., 175.

Qualls, R.G., *et al.* (1983) The Role of Suspended Particles in Ultraviolet Disinfection. *J. Water Pollut. Control Fed.*, **55**, 1280.

Rein, D.A., *et al.* (1992) Toxicity Effects of Alternate Disinfection Processes. Paper presented at 65th Annu. Conf. Water Environ. Fed., New Orleans, La.

Scheible, O.K. (1987) Development of a Rationally Based Design Protocol for the Ultraviolet Light Disinfection Process. *J. Water Pollut. Control Fed.*, **59**, 1, 25.

Severin, B.F., *et al.* (1983) Kinetic Modeling of U.V. Disinfection of Water. *Water Res. (G.B.),* **17**, 1669.

Severin, B.F., *et al.* (1984) Mixing Effects in U.V. Disinfection. *J. Water Pollut. Control Fed.*, **56**, 881.

U.S. Environmental Protection Agency (1986a) *Design Manual: Municipal Wastewater Disinfection.* EPA-625/1-86-021, Water Eng. Res. Lab., Cincinnati, Ohio.

U.S. Environmental Protection Agency (1986b) *Quality Criteria for Water.* EPA-440/5-86-001, Office Regulations Stand., Washington, D.C.

U.S. Environmental Protection Agency (1992) *Ultraviolet Disinfection Technology Assessment.* EPA-832/R-92-004, Office of Water, Washington, D.C.

Water Environment Federation (1993) *Proc. Plann., Des. & Oper. Effluent Disinfection Syst. Spec. Conf.,* Whippany, N.J.

Whitby, G.E., and Palmateer, G. (1993) The Effects of UV Transmission, Suspended Solids, Wastewater Mixtures and Photoreactivation in Wastewater Treated with UV Light. *Proc. Plann., Des. & Oper. Effluent Disinfection Syst. Spec. Conf.,* Water Environ. Fed., Whippany, N.J., 24.

White, G.C. (1992) *The Handbook of Chlorination and Alternative Disinfectants.* 3rd Ed., Van Nostrand Reinhold, New York, N.Y.

Wilson, B., *et al.* (1992) Coliphage MS-2 as UV Water Disinfection Efficacy Test Surrogate for Bacterial and Viral Pathogens. Poster presented at Water Qual. Technol. Conf. Am. Water Works Assoc.

Wolfe, R.L. (1990) Ultraviolet Disinfection of Potable Water. *Environ. Sci. Technol.,* 24, 768.

Yip, R.W., and Konasewich, D.E. (1972) Ultraviolet Sterilization of Water—Its Potential and Limitations. *Water Pollut. Control*, **14**, 14.

Zubrilin, N.G., *et al.* (1991) Combined Effect of Krypton Monofluoride Laser Radiation and Copper Ions on the Survival Rate of *E. coli* Cells. *Khim. Khim. Tekhnol.* (Kiev), **13**, 362.

Index

A

Acid chrome violet K, 167
Acquired immunodeficiency syndrome (AIDS), 37
Activated-sludge system, pure-oxygen, 159
Aftergrowths, 113
Air compressors, size and number, 205
Amperometry, 142, 167
  endpoint, 130
  method, 128, 129
Arrhenius equations, 49
Axial dispersion number, 77

B

Bacillus subtilis, 248, 257
  spores, 280
Bacteria, 31
  densities in domestic wastewater, 38
  heterotrophic, 248
Band control, 145
Beer's law, 237, 238, 250
Bench-scale tests, 180
Bernoulli principle, 135
Bioassay, 258, 259
  method, 257
BOCA National Fire Code, 120

C

Calorimetry, 165
  procedure, 166
Canada
  disinfection requirements, 15
  Federal-Provincial-Territorial partnership, 14
  legislation, 13, 14
  maximum coliform limits, 15
  regulations and guidelines, 14
Candida parapsilosis, 244
Carbon, 117
  granular, 117
  powdered, 117
Carcinogens, 10, 52
Cascade control, 145, 150
Catalytic destruction, 198
Chemical-feed
  pumps, 136
  system, 138
Chemiluminescence, 165, 166
Chick-Watson, 173
  equation, 90, 94
  model, 50, 172
Chick's law, 49, 50
Chloramines, 109
Chlorination
  breakpoint, 110
  free, 116
Chlorinators, 135
  flow proportional, 146

Chlorine, 45, 106, 110, 114
    chemistry of, 102, 104, 106
    combined, 48
    commercial, 102
    contact chamber, 97
    containers, 121
    cylinders, storage, 119–121
    design requirements, 132
    diffuser, 133
    dioxide, 107
    effects on higher organisms, 113
    elemental, 102
    feed rate, 144, 146
    free, 132
    gas, 117, 135, 144
    gas detectors, 121
    hypochlorous acid, 108
    liquid, 125
    physical properties, 102
    requirement, 131
    solubility, 102
    storage, 119, 121
    toxicity, 104, 113
    vaporizers, 140, 141
Chlorine residual, 116, 128, 129,
    130, 153, 170
    analytical determination, 128
    analyzer, 153
    combined, 114
    free, 114, 116
    reduction, 116
Chlorine-contacting device, 133
Closed conduits, 132
Closed-loop control, 144
Coliform
    bacteria, 7, 39, 173, 174, 241, 248
    bacterial densities, 38
    fecal, maximum limits in Canada,
        15
    collimated beam, experiments,
        238
Collins model, 51
Collins-Selleck model, 247, 248

Colorimetry, 129
Compound-loop control, 144, 149
Concentration time (CT), 172, 173
Contact time, 48
Contacting, 133, 188
Contactor design considerations,
    195
Contactors, 191
Continuously stirred tank reactor, 63
Cost
    installation, 287
    maintenance, 288
    operating, 276
    relamping, 288
    savings, 287
Cryptosporidium, 26, 177, 178
Cylinders, gas, 119

D

Dark repair, 233
Dechlorination
    control, 153
    feed-back schematic, 153
    feed-forward schematic, 154
    reactions, 116
Desiccant dryers, size and number,
    206
Design, 276
    contactors, 195
    facilities, 123
    feasibility, 179
    handling facilities, 121
    reactors, 97
    retrofit considerations, 282
    ultraviolet disinfection systems,
        280
Direct contact route, 29
Direct reading analyzer, 154
Disease, 28, 32
    fundamentals, 28
    recreational water, 26
    vectors, 28

Disinfectants
    alternatives, 42
    residual toxicity, 52
Dispersion model, *E* curve, 78
Drinking water, 21, 45

E

Effluent, 276
    coliform viability, 239, 271, 272
    fecal coliform concentrations,
        278
    virus concentrations, 114
Electrical power supply, 210, 212
Electromagnetic radiation, energy,
    230
Energy loss, 264
Enteric bacteria, 7
Epidemics, 21, 28
    studies, 28
    terms, 28
Equipment design and selection, 135
Escherichia coli, 11
    kill times versus residual
        concentration, 115
    survival, 176
Etiologic agents, 235
Evaporators, 124

F

Feed-back control, 144
    system, 153
Feed control strategies, 144
Feed-forward control, 144, 154
Feed gas
    moisture content, 183
    preparation, 181
    selection, 181
Fine-bubble diffusers, 195
Flow pacing, 144

Flow-proportional control, 144, 146,
    147
Fouling, 273
    lamp, 275
Free chlorine residual, 114

G

Gas
    feed rate, 144
    induction system, 195
    mass transfer, 189
    preparation, 209, 211
    treatment systems, 182
Gas-flow rate, 188
Gas-phase concentration
    measurement, 165

H

Hazen theorem, 27
Head loss, 264, 266
Health, 117
Human immunodeficiency virus
    (HIV), 37
Hydraulics, 259
    design, 280
    devices, 133
Hypochlorites, 104, 118
    chemical properties, 105
    physical properties, 105

I

Immunization, 30
Inactivation behavior, effect of
    intensity on, 248
Inactivation kinetics, 246
Indicator organisms, 9, 37

Infections, 29
  preventive measures, 29, 30
  routes of, 29
  spread of, 29
Infective dose, 28
Injector and static mixer, 197
Inorganic reactions, 109
Iodometry, 128, 130, 165, 166
  method, 165, 166
  procedure, 167

K

Kinetics, 48, 244
  disinfection, 48, 90
  models, 48, 244, 247
  series-event, 247
  ultraviolet disinfection,
    inactivation, 236

L

Lag time, 147, 148
Lamps
  cleaning, 285, 286
  output, 267
Leuco crystal violet, 129
Longitudinal dispersion, 259, 267

M

Maintenance of ozonation systems,
  208, 211
Manifolds, location, 139
Manual control, 144
Manual injection points, 150
Manual sampling, 148
Microbial
  fundamentals, 28
  inactivation, 45, 47, 52, 245

Microbiological systems, 232
Midpoint method, 70
Mills-Reincke phenomenon, 27
Mixing, chlorine, 131
Morril dispersion index, 81

N

N,N-diethyl-*p*-phenylene-diamine,
  129
N,N-dimethyl analine, 129
Nonideal
  conceptual models, 76
  dispersion model, 77
  reactor theory, 269
  reactors, 75
  segregation model, 77
  tanks-in-series model, 79

O

Offgas destruction, 198
  system, 210
On-line chlorine residual analyzer,
  142
Open-loop control, 144
Organic reactions, 111
Otto plate, 185
Oxidation-reduction potential, 52
Ozonation, 209
  feed gas selection and
    preparation, 181
  process components, 188
  source gas, 181
Ozone, 45
  aqueous-phase concentration
    measurement, 167
  chemistry in aqueous solution,
    161
  concentration, 165, 166, 186
  contactor, 210, 212

cooling, 186
decomposition, 161, 162, 172
demand, 165
design production rate, 203
disinfection, 174
dose response curve, 202
effect on bacteria, 175
effect on protozoa, 177
effect on viruses, 175
enthalpy of decomposition, 166
fate of, 161, 163
generation, 184
generator, 186, 203, 210, 212
mass transfer, 50
oxidation, 172
production, 184
properties, 159, 160
residual toxicity, 170
toxicological properties, 168
transfer efficiency, 94, 203

P

Pathogens, 28, 37
    fundamentals, 28
    survival in the environment, 39
    types, 30
Pearson correlation coefficient, 11
Peclet number, 78
Photobiochemical
    change, 236
    inactivation, 246
    reaction, 245
Photoreactivation, 46, 233, 234
Pilot testing, 180, 279
Plug-flow reactor, 62
Point source summation, 259
    method, 258
Process control, 197
    design example, 198
    design requirement, 131
Public health concerns, 7

Q

Quartz jacket fouling, 273
Quartz sleeve fouling, 279

R

Reaction pathways, 161
Reactors, 75
    baffles, 97
    configuration, 97
    design considerations, 97
    efficiency, 92, 93
Recreational water, disease
    outbreaks, 26
Regrowth phenomenon, 51
Regulatory concerns, 9
    Canada, 12, 14
    national, 9
    state, 10
Renton system, 153
Residence time distribution, 69
Residual
    analyzers, 142
    control, 144, 147
    free, 116
    toxicity of disinfectants, 52
Retrofit considerations, 282

S

Safety, 117, 208, 213
    chlorine gas, 117
    cylinders, 119
    hypochlorites, 118
    shipment and handling, 119
    storage, 119
    sulfur dioxide, 118
Scheible model, 269, 270
Secondary effluent, 241, 243

Semiautomatic control, 145
Series-event kinetic model, 247
Serpentine flow reactors, 75, 76
Severin model, 273
Shellfish, 27
Staehelin, Bader, and Hoigné,
   mechanism, 161
Starch-iodide, 129
   method, 128
Sulfonators, 135
Sulfur chemistry, 102
Sulfur dioxide, 118
   chemical properties, 106
   containers, 121
   equipment, 124
   feeders, 138
   gas detectors, 121
   physical properties, 105
   storage, 119, 121
   toxicity, 106
   vaporizers, 140, 141
Suspended solids, 276, 277
   concentrations, 278
   data, 279
Syringaldazin, 130

T

Titration, 129
Tomiyasu, Fukutomi, and Gordon
   mechanism, 161
Ton containers, 124
   manifold, 126
   orientation, 123
Total residual chlorine
   concentrations, 15
Tracer analysis
   disinfection kinetics, 88, 90
   disinfection reactors, 97
   dispersion model, 92
   examples, 82
   ideal models, 91
   pulse-feed test, 84

   segregated model, 91
   step-feed test, 82
   tanks-in-series model, 92
Tracer curve, 69
   C curve, 71
   curve, variance, 74
   E curve, 69, 70
   F curve, 71
   mean residence time, 73
Tracer data
   curve parameters, 81
   d and N indices, 80
   index of average detention , 81
   index of initial short-circuiting,
     81
   index of mean detention, 81
   index of modal detention time, 81
   interpretation, 80
   other index values, 80
Tracers,
   detection, 63, 67
   injection, 63
   pulse input, 67–69
   selection, 66
   step input, 67, 68
Trihalomethanes, 112
Turbines, 197

U

Ultraviolet disinfection, 230
   absorbance, 165, 251
   absorption, 167
   cleaning, 282
   continuous flow, 249
   design, 280
   dissipation mechanism, 250
   dose, 257
   equipment, 283
   horizontal, 284
   inactivation kinetics, 236
   intensity, 249, 252, 253
   intensity distributions, 254

light, 41
low-pressure, high-intensity, 288
low-pressure mercury lamp, 283
mathematical models, 269
medium-pressure mercury lamp,
    287, 288
open-channel, 283
output, 267
radiation, 46
transmittance, 276
vertical, 285

V

Vacuum regulator, location, 139

Vapor pressure, 141
Vaporizers, 140, 141
    facilities, 124

W

Water quality
    disinfection criteria, 20
    standards, 20
Waterborne disease, 8, 21, 36
    outbreaks, 21
    transmission routes, 8